6th Grade Nevada Math for Beginners

Standardized Testing and Home School Study Guide

MATHFA

www.mathnotion.com

… So Much More Online!

✓ FREE Mathematics Worksheets

✓ More Math Learning Books!

✓ FREE Math Lessons

✓ Online Math Tutors

✓ **PDF Version of This Book**

No Registration Required!

Visit us here!

6th Grade Nevada Math for Beginners

By Mathfa, Email: info@mathfa.com

Web: www.mathnotion.com

Copyright © 2025 by Mathfa, an imprint of Math Notion Inc. All rights reserved. No part of this publication may be reproduced, stored in a retrieval system, or transmitted in any form or by any means, electronic, mechanical, photocopying, recording, scanning, or otherwise, except as permitted under Section 107 or 108 of the 1976 United States Copyright Act, without permission of the author.

All inquiries should be addressed to Mathfa.

ISBN: 978-1-63620-255-6

Welcome to Math Grade 6!

This book is designed to help you build a solid foundation in math while preparing you for more advanced topics in the years ahead. Whether you are just getting comfortable with new concepts or looking for extra challenges, this book has something for everyone.

Inside, you will find a wide range of topics, from fractions, decimals, and ratios to equations, geometry, and statistics. Each chapter is carefully structured to help you progress step by step, reinforcing what you have learned while introducing innovative ideas in a clear and engaging way.

This book is not just for one type of student, it is designed for all learners. If you need extra support, you will find guided examples and practice problems. If you are looking for a challenge, special sections marked for higher-level thinking will push you to explore deeper and think critically.

Throughout the book, you will see how math connects to the real world. With word problems, activities, and real-life applications, you will understand not just how math works, but why it matters. Plus, there are more worksheets and exercises to help you practice and sharpen your skills.

Whether you are a student working independently, a teacher guiding a class, or a parent supporting a learner, this book is here to make math accessible, engaging, and even fun.

www.mathnotion.com

How to effectively use the book?

Using the step-by-step guidebook, like one for Math Grade 6, effectively can make a significant difference in understanding and mastering the subject. Here are steps and strategies to get the most out of your math study guidebook:

1) **Familiarize Yourself with the Contents:**
   - Scan through the entire book to get an overview of the topics covered.
   - Take note of the structure. Are there chapters, sections, examples, and solutions?

2) **Set Clear Goals:**
   - Determine what you want to achieve. Is it a better understanding of a particular topic, exam preparation, or just general knowledge enrichment?
   - Break your goals down into smaller, manageable tasks.

3) **Plan Your Study Time:**
   - Allocate specific times for studying. Consistency can help reinforce learning.
   - Avoid cramming. Spread your study sessions out over time.

4) **Active Reading:**
   - Do not just passively read. Engage with the content.
   - Highlight key points, make annotations, and jot down summaries.

5) **Work Through Practice Problems:**
   - The essence of Math is problem-solving. Make sure to work through the practice problems in the guidebook.
   - After attempting problems, check the solutions. If you get something wrong, understand why.

6) **Use Additional Resources:**
   - If there is a concept you are struggling with, seek additional resources like online videos, tutorials, or other textbooks. Sometimes a different explanation can make things click.

7) **Review Regularly:**
   - Periodic review helps reinforce what you have learned.
   - Revisit challenging topics or problems you initially got wrong.

www.mathnotion.com

8) Study Actively and Engage Multiple Senses:

- Explain difficult concepts aloud as if you are teaching someone else.
- Draw diagrams or use physical objects to represent algebraic concepts.

9) Join a Study Group:

- Discussing and teaching topics can solidify your understanding.
- Other students might have insights or understanding that you have not considered.

10) Seek Help When Needed:

- Do not be afraid to ask for help if you are stuck.
- This could be from a teacher, tutor, classmate, or online forum.

11) Evaluate Yourself:

- If your guidebook has quizzes or mock exams, take them seriously.
- This helps gauge your understanding and readiness for any actual exams.

12) Stay Curious and Relate to Real Life:

- Try to relate algebraic concepts to real-world applications. This not only makes it more interesting but can also aid understanding.

Remember, the key is consistency and active engagement. With regular, focused effort and the right strategies, you can use your 6th Grade Math study guide to its full potential and master the subject.

Table of Contents

Chapter 1: Whole Numbers Review 13

Adding and Subtracting Whole Numbers 14
Multiplying Whole Numbers 15
Division with 2-Digit and 3-Digit Divisors 16
Division with Remainders 17
Numbers Ending in Zeros 18
Rounding Techniques 19
Estimation Techniques 20
Word Problems 21
Worksheets 22
Answer of Worksheets 25

Chapter 2: Integers 27

Integers and Opposite Integers 28
Integers and Absolute Value 29
Ordering Integers and Numbers 30
Adding and Subtracting Integers 31
Multiplying and Dividing Integers 32
Order of Operations 33
Word Problems 34
Worksheets 35
Answer of Worksheets 38

Chapter 3: Number Theory 39

Prime and Composite Numbers 40
Divisibility Rules 41
Identify Factors 42
Prime Factorization 43
Greatest Common Factor 44
Least Common Multiple 45
Word Problems 46
Worksheets 47
Answer of Worksheets 49

Chapter 4: Fractions 51

Fractions and Mixed Numbers 52

www.mathnotion.com

Ordering Fractions 53
Simplifying Fractions 54
Least Common Denominator 55
Adding and Subtracting 56
Multiplying and Dividing 57
Mixed Operations 58
Word Problems 59
Worksheets 60
Answer of Worksheets 63

Chapter 5: Decimals 65

Ordering Decimals 66
Fraction and Decimal 67
Mixed Numbers and Decimals 68
Adding and Subtracting Decimals 69
Multiplying Decimals 70
Dividing Decimals by whole numbers 71
Dividing Decimals by Decimals 72
Multiply and Divide Decimals by the Power of Ten 73
Rounding Decimals 74
Word Problems 75
Worksheets 76
Answer of Worksheets 79

Chapter 6: Ratios, Proportions and Percents 81

Writing and Identifying Ratios 82
Simplifying Ratios 83
Equivalent Ratios and Rates 84
Proportion in Tables and Graphs 85
Ratio and Rates Word Problems 86
Converting Between Fractions, Mixed Numbers and Percent 87
Converting Between Percent and Decimals 88
Finding Percents of Numbers 89
Discount, Tax and Tip 90
Word Problems 91
Worksheets 92
Answer of Worksheets 95

Chapter 7: Exponents and Radicals Expression 97

Evaluating Powers 98
Finding Missing Exponents 99
Adding and Subtracting Exponents 100
Multiplication Property of Exponents 101

www.mathnotion.com

Division Property of Exponents 102
Negative Exponents and Negative Bases 103
Decimal and Fractional Bases 104
Powers of Ten 105
Scientific Notation 106
Square Roots 107
Word problems 108
Worksheets 109
Answer of Worksheets 112

Chapter 8: Measurements 115

Reference Measurement 116
Metric Length Measurement 117
Customary Length Measurement 118
Metric Capacity Measurement 119
Customary Capacity Measurement 120
Metric Weight and Mass Measurement 121
Customary Weight and Mass Measurement 122
Temperature 123
Time 124
Word Problems 125
Worksheets 126
Answer of Worksheets 129

Chapter 9: Algebraic Expressions 131

Find a Rule 132
Writing Variable Expressions 133
Identify Terms and Coefficients 134
Simplifying Variable Expressions 134
Equivalent Expressions - Properties 135
Evaluating Variable Expressions 136
Equivalent Expressions - Strip Model 137
Factor Numerical Expressions 138
Factor Variable Expressions 139
Factor Variable Expressions - Area Model 140
Word problems 141
Worksheets 142
Answer of Worksheets 145

Chapter 10: Equations and Inequalities 147

Identify Expressions and Equations 148
One–Step Equations 149
Two-Step Equations 150

www.mathnotion.com

Multi-Step Equations 151
Equation Diagram 152
One-Step Inequalities 153
Two-Steps Inequalities 154
Multi-Step Inequalities 155
Graphing Inequalities 156
Word Problems 157
Worksheets 158
Answer of Worksheets 162

Chapter 11: Two Variable Equations 165

Describe Coordinate Plane 166
Directions on Coordinate Plane 167
Reflect a Point Over an Axis 168
Distance Between Two Points 169
Identify Independent and Dependent Variables 170
Area and Perimeter of Shape on the Coordinate Plane 171
Write an equation from a Graph Using a Table 172
Value of Two Variables Equations 173
Word Problems 174
Worksheets 175
Answer of Worksheets 180

Chapter 12: Geometry and Solid Figures 182

Lines, Segments, and Angles 183
Complementary and Supplementary Angles 184
Symmetry and Congruence 185
Classifying Polygons 186
Classifying Triangles 187
Pythagorean Relationship 188
Parallelograms 189
Rectangles 190
Triangles 191
Trapezoids 192
Circles 193
Area of Compound Shapes 194
Cubes 195
Rectangular Prisms 196
Triangular Prisms 197
Cylinder 198
Word Problems 199
Worksheets 200
Answer of Worksheets 209

www.mathnotion.com

Chapter 13: Statistics and Probabilities 213

Mean and Median 214
Mode and Range 215
Histograms 216
Line Graphs 217
Times Series 218
Stem -and -Leaf Plot 219
Quartile of a Data Set 220
Box- and -Whisker Plots 221
Dot Plots 222
Pie Graph 223
Counting principle 224
Probability of Simple Events 225
Probability of opposite events 226
Word Problems 227
Worksheets 228
Answer of Worksheets 233

Chapter 14: Practice Test 239

Formula Sheet 240
SBAC Practice Test 241
Answer Key 246
Answers and Explanations 247

Chapter 1: Whole Numbers Review

Topics that you will learn in this chapter:

- Adding and Subtracting Whole numbers
- Multiplying Whole Numbers
- Division with 2-Digit and 3-Digit Divisors
- Division with Remainders
- Numbers Ending in Zero
- Rounding Techniques
- Estimation Techniques
- Word Problems
- Worksheets

Adding and Subtracting Whole Numbers

Adding and subtracting whole numbers is like counting and reversing counting. When you add two whole numbers, you simply combine their values and when you subtract one whole number from another, you remove the value of the second number from the first. You should note that for more than one-digit numbers you must go through the following steps:

a) Line up the numbers vertically by place value.

b) Add from right to left (start with the ones place).

c) **For adding**: If a column adds up to more than 9, carry over the extra value to the next column on the left.

d) **For subtracting**: if the top number is smaller than the bottom number in a column, borrow from the next column to the left.

Examples:

1) Add the two numbers 2,769 and 3,593. **Solution:**

   Step 1: Write the numbers vertically.

   $$\begin{array}{r}2,769\\+3,593\\\hline\end{array}$$

   Step 2: Add the rightmost digits first (units place)

   $9 + 3 = 12$

   - Write down 2 and carry over 1

   $$\begin{array}{r}2,769\\+3,593\\\hline 2\end{array}$$

   Step 3: Add the tens place: $6 + 9 = 15$

   - Add the carried over 1 to get 16.
   - Write down 6 and carry over 1.

   $$\begin{array}{r}2,769\\+3,593\\\hline 62\end{array}$$

   Step 4: Add the hundreds place: $7 + 5 = 12$

   - Add the carried over 1 to get 13.
   - Write down 3 and carry over 1.

   $$\begin{array}{r}2,769\\+3,593\\\hline 362\end{array}$$

   Step 5: Add the thousands place: $2 + 3 = 5$

   - Add the carried over 1 to get 6.

   $$\begin{array}{r}2,769\\+3,593\\\hline 6,362\end{array}$$

2) Subtract two numbers 8,059 and 7,234. **Solution:**

   Step 1: Write the numbers vertically

   $$\begin{array}{r}8,059\\-\ 7,234\\\hline\end{array}$$

   Step 2: Subtract the rightmost digits (units place):

   $$\begin{array}{r}8,059\\-7,234\\\hline 5\end{array}$$ $9 - 4 = 5$

   Step 3: Subtract the tens place: $5 - 3 = 2$

   $$\begin{array}{r}8,059\\-\ 7,234\\\hline 25\end{array}$$

   Step 4: Subtract the hundreds place: $0 - 2$, since we can't subtract 2 from 0, we need to borrow 1 from the thousands place: $10 - 2 = 8$, so adjust the thousand places $7 - 7 = 0$

   $$\begin{array}{r}8,059\\-\ 7,234\\\hline 825\end{array}$$

www.mathnotion.com

Multiplying Whole Numbers

There is a standard algorithm for multiplying whole numbers that can be followed as below:

a) Write the numbers vertically, one below the other.
b) Multiply each digit of the top number by each digit of the bottom number, starting from the rightmost digit.
c) Carry over any extra value to the next column, if necessary.
d) Add the results of each step.

Examples:

1) Find the product of the two numbers 189 and 37:

   Solution:

   Step 1: Write the numbers vertically.

   $$\begin{array}{r} 189 \\ \times 37 \\ \hline \end{array}$$

   Step 2: Multiply 189 by 7 (the units digit of 37):
   - When we multiply 7 by 9 the result is 63, so we carry 6 to the next column to add it up by 56 (8 ×7) and then we repeat it for the last column:

   $$\begin{array}{r} 189 \\ \times 37 \\ \hline 1{,}323 \end{array}$$

   Step 3: Multiply 189 by 30 (the tens digit of 37, where 30 = 3 ×10) and write it under the 1,323:
   - Again, we must carry over any extra value to the next column as the previous step:

   $$\begin{array}{r} 189 \\ \times 37 \\ \hline 1{,}323 \\ 5{,}670 \\ \hline \end{array}$$

   Step 4: Add the results of two previous steps:

   $$\begin{array}{r} 189 \\ \times 37 \\ \hline 1{,}323 \\ +5{,}670 \\ \hline 6{,}993 \end{array}$$

2) Find the product of the two numbers 9,002 and 241:

   Solution:

   The explanation for each step is just like in the previous example!

   $$\begin{array}{r} 9{,}002 \\ \times 241 \\ \hline 9{,}002 \end{array} \rightarrow \begin{array}{r} 9{,}002 \\ \times 241 \\ \hline 9{,}002 \\ 360{,}080 \end{array} \rightarrow \begin{array}{r} 9{,}002 \\ \times 241 \\ \hline 9{,}002 \\ 360{,}080 \\ +1{,}800{,}400 \\ \hline 2{,}169{,}482 \end{array}$$

Division with 2-Digit and 3-Digit Divisors

Dividing by 2-digit and 3-digit numbers involves long division and you must follow the steps below:

a) Take the first digit of the dividend from the left.
b) Then divide it by the divisor and write the answer on top as the quotient.
c) Subtract the result from the digit and write the difference below.
d) Bring down the next digit of the dividend.
e) Repeat the same process.
☑ At each step, if the resulting number is smaller than the divisor after lowering the next digit, we can move the next digit down after putting zero in quotient part.

Examples:

1) Find the quotient of 779 by 41:

   Solution:

   Step 1: Write 779 under the division bar and 41 outside:

   $41\overline{)779}$

   Step 2: look at the first two digits of 779 (77). How many times does 41 go to 79? It goes in one time, write 1 above the division bar:

   $$\begin{array}{r} 1 \\ 41\overline{)779} \\ -41 \\ \hline 36 \end{array}$$

   Step 3: Bring down the 9 from 779 to make 369:

   $$\begin{array}{r} 1 \\ 41\overline{)779} \\ -41 \\ \hline 369 \end{array}$$

   Step 4: how many does 41 go into 396? It goes in 9 times. 41 multiplied by 9 equals 396 and finally subtract 396 from 396 to get 0:

   $$\begin{array}{r} 19 \\ 41\overline{)779} \\ -41 \\ \hline 369 \\ -369 \\ \hline 000 \end{array}$$

2) Find the quotient of 13,823 by 23. **Solution:**

   You should note that in the second step when you put down the 2, since it is lower than 23, first we put a zero beside the quotient (6) and then put down the next digit (3) beside the 2:

   $$\begin{array}{r} 6 \\ 23\overline{)13,823} \\ -138 \\ \hline 000 \end{array} \rightarrow \begin{array}{r} 60 \\ 23\overline{)13,823} \\ -138 \downarrow \\ \hline 0002 \end{array} \rightarrow \begin{array}{r} 601 \\ 23\overline{)13,823} \\ -138 \downarrow \\ \hline 00,023 \\ -23 \\ \hline 00 \end{array}$$

www.mathnotion.com

Division with Remainders

The process of division with remainders is almost the same as the previous topic:

a) Starting from the leftmost digit of the dividend, determine how many times the divisor can fit into that part of the dividend.
b) Write the result (quotient) above the division bar.
c) Multiply the quotient by the divisor, subtract this result from the relevant part of the dividend, and write the remainder below the subtraction line.
d) Bring down the next digit of the dividend next to the remainder.
e) Repeat the divide, multiply, subtract, and bring down process until there are no more digits to bring down.
f) If there is any number left that is smaller than the divisor after all digits have been brought down, this number is the remainder.

Example:

✱ Find the quotient of 89,057 by 419:

Solution:

Step 1: Write 89,057 under the division bar and 419 outside:

$$419 \overline{)89{,}057}$$

Step 2: Divide the first few digits (890) by 419: 419 goes into 890 two times:

$$\begin{array}{r} 2 \\ 419\overline{)89{,}057} \\ -838 \\ \hline 52 \end{array}$$

Step 3: Bring down the 5 from 89,057 to make 525:

$$\begin{array}{r} 2 \\ 419\overline{)89{,}057} \\ -838 \\ \hline 525 \end{array}$$

Step 4: How many does 419 go into 525? It goes in 1 time. Then bring down the 7 to make 1,067:

$$\begin{array}{r} 21 \\ 419\overline{)89{,}057} \\ -838 \\ \hline 525 \\ -419 \\ \hline 1{,}067 \end{array}$$

Step 5: 419 goes into 1,067, 2 times:

$$\begin{array}{r} 212 \\ 419\overline{)89{,}057} \\ -838 \\ \hline 525 \\ -419 \\ \hline 1{,}067 \\ -838 \\ \hline \end{array}$$

Step 6: 419 times 2 is 838. Subtract 838 from 1067 to get 229. So, the remainder is 229:

$$\begin{array}{r} 212 \\ 419\overline{)89{,}057} \\ -838 \\ \hline 525 \\ -419 \\ \hline 1{,}067 \\ -838 \\ \hline 229 \end{array}$$

Numbers Ending in Zeros

Multiplying Numbers Ending in Zeros:

a) Ignore the zeros and multiply the non-zero digits.
b) Count the total number of zeros at the end of both numbers.
c) Attach the zeros back to the product.

Dividing Numbers Ending in Zeros:

a) Cancel out the same number of zeros from the dividend and the divisor.
b) Divide the simplified numbers.
c) If there are more zeros in the dividend than the divisor, attach the remaining zeros to the quotient.
☑ You must note that this technique cannot be used when the divisor is not a factor of the dividend, because this kind of division contains the remainder.

Examples:

1) Multiply the numbers 2,000 and 300 together:
   Solution:
   Step 1: Multiply the non-zero parts: $2 \times 3 = 6$.

   Step 2: Count the zeros: 2,000 has 3 zeros, 300 has 2 zeros. Total: 5 zeros.

   Step 3: Attach the zeros: 6 followed by 5 zeros is 600,000.

2) Divide the numbers 80,000 and 400:
   Solution:
   Step 1: Remove two zeros from each: 80,000 becomes 800, 400 becomes 4.
   Step 2: Divide the simplified numbers: $800 \div 4 = 200$.
   Step 3: No extra zeros to attach. So, $80,000 \div 400 = 200$.

3) Find the missing number in each expression:
   a) $2,000 \times \square = 600,000$ b) $\square \div 500 = 3$ c) $\square \times 120 = 24,000$
   Solution:
   We should note that the inverse of multiplication is division, and vice versa!
   a) To find the missing number, we need to divide 600,000 by 2,000. So $600,000 \div 200 = 3,000$.
   b) Just like part a, you need to do the reverse operation, which is multiplication. So, $500 \times 3 = 1,500$.
   c) This expression is similar to part a: $24,000 \div 120 = 200$.

Rounding Techniques

Rounding whole numbers is straightforward once you know the rules:

a) **Identify Place Value**: Decide which place you want to round to (nearest ten, hundred, etc.).

b) **Look at the Digit to the Right:** Check the digit immediately to the right of the place you're rounding to.

c) **Round Up or Down:**
   - If the digit is 5 or greater, round up by increasing the rounding digit by 1.
   - If the digit is less than 5, round down by keeping the rounding digit the same.

Examples:

1) Round 276 to the nearest ten:

   Solution:

   Step1: Identify the place value: Ten place (7 in 276).

   Step 2: look at the digit to the right: It is 6.

   Step 3: Round Up: Because 6 is 5 or greater, increase the tens place by 1 (7 becomes 8).

   So, 276 rounded to the nearest ten is 280.

2) Round 987,654 to the nearest hundred thousand.

   Solution:

   Step 1: Identify the place value: We need to round the nearest hundred thousand. The number in the hundred thousand places is 9.

   Step 2: Look at the digit to the right: The digit immediately to the right of 9 is 8.

   Step 3: Round Up: Since 8 is greater than 5, we round up.

   Step 4: So, 987,654 rounded to the nearest hundred thousand is 1,000,000.

1) Round each number to the underlined place value:

   a) 5̲5   b) 2̲48   c) 4̲70,886   d) 8̲,319   e) 19,9̲99

   Solution:

   a) The digit immediately to place value (5) is 5 and we round up by increasing the digit 5 by 1, so 55 ≈ 60.

   b) The digit immediately to place value (2) is 4 and we round down by keeping digit 2 the same, so 248 ≈ 200.

   c) the digit immediately to place value (4) is 7 and we round up by increasing the digit 4 by 1, so 470,886 ≈ 500,000.

   d) The digit immediately to place value (8) is 3 and we round down by keeping the digit 8 the same, so 8,319 ≈ 8,000.

   e) the digit immediately to place value (9) is 9 and we round up by increasing the digit 9 by 1, so 19,999 ≈ 20,000.

www.mathnotion.com

Estimation Techniques

Estimation techniques for whole numbers help you quickly get an approximate answer without needing exact calculations. Three of the most important estimation techniques are as follows:

- **Rounding:** Round numbers to the nearest ten, hundred, thousand etc., to make calculation easier.
- **Front-End Estimation:** Use the leading digit and ignore the smaller ones.
- **Compatible Numbers:** Adjust numbers to ones that are easier to work with mentally.

Examples:

1) Estimate the sum and subtraction by rounding each number to the nearest hundred (use the *rounding technique*):

   a) $1,582 - 391 =$ b) $6,792 + 5,894 =$ c) $74,011 - 4,390 =$

   Solution:

   Based on rounding rules:

   a) $1,582 \approx 1,600$, $391 \approx 400 \to 1,600 - 400 = 1,200$.
   b) $6,792 \approx 6,800$, $5,894 \approx 5,900 \to 6,800 + 5,900 = 12,700$.
   c) $74,011 \approx 74,000$, $4,390 \approx 4,400 \to 74,000 - 4,400 = 69,600$.

2) Estimate the product of the following expressions by rounding each number to the nearest thousand (use the *front-end estimation* technique):

   a) $1,489 \times 2,840 =$ b) $158 \times 18,572 =$ c) $2,413 \times 950,214 =$

   Solution:

   a) $1,489 \approx 1,000$, $2,840 \approx 2,000 \to 1,000 \times 2,000 = 2,000,000$.
   b) $0158 \approx 0$, $18,572 \approx 18,000 \to 0 \times 18,000 = 0$.
   c) $2,413 \approx 2,000$, $950,214 \approx 950,000 \to 2,000 \times 950,000 = 1,900,000,000$

3) Estimate division expressions below to make them easier to work:

   a) $521 \div 11 =$ b) $1,829 \div 823 =$ c) $57,102 \div 68 =$

   Solution:

   In this case we should use the *compatible number* technique:

   a) $521 \approx 500$, $11 \approx 10 \to 500 \div 10 = 50$.
   b) $1,829 \approx 1800$, $823 \approx 900 \to 1,800 \div 900 = 2$.
   c) $57,102 \approx 56,000$, $68 \approx 70 \to 56,000 \div 70 = 800$

Word Problems

Solving word problems requires a bit of translating the problem from words into math. Here's a step-by-step approach to tackle them:

a) **Read the Problem Carefully:** Understand what the problem is asking. Identify the key information and what you need to find out.

b) **Underline or Highlight Important Information:** Underline or highlight numbers, operations, and keywords that hint at what kind of math operation you'll need.

c) **Translate Words into a Math Equation**: Turn the words into a mathematical expression or equation. Identify operations like addition, subtraction, multiplication, or division.

d) **Solve the Math Equation:** Do math! Solve the equation using the appropriate operations.

e) **Check Your Work**: Review your steps and make sure your answer makes sense in the context of the problem.

Examples:

1) David had $398.80 to buy some Christmas gifts, he bought perfume for his sister with $109.34 and a pair of shoes for his mother for $89.99, how many dollars did he have left?

Solution:

Step 1: Read the problem carefully: David starts with $398.80 and buys two gifts.

Step 2: Underline or highlight important information:

- Initial amount: $398.80.
- Cost of perfume: $109.34.
- Cost of shoes: $89.99.

Step 3: Translate words into math equations:

Remaining amount = 398.80 – (109.34 + 89.99).

Step 4: Solve the math equation:

```
 109.34          398.80
+ 89.99         −199.33
 199.33          199.47
```

Step 5: Check your work.

David has $199.47 left after buying the gifts.

www.mathnotion.com

Worksheets

Adding and subtracting whole number

Do the following additions and subtraction operations:

1) 7,009 + 4,992

2) 8,520 + 6,483

3) 15,999 + 2,615

4) 10,769 − 8,872

5) 29,512 − 4,411

6) 90,000 − 86,117

Find the missing number:

7) 1,257 − .... = 856

8) .... − 1,098 = 811

9) 124 + ... = 7,001

10) .... + 2,984 = 10,000

Multiplying whole numbers

Find the product of the following multiplication expressions:

1) 85 × 29

2) 103 × 9

3) 289 × 77

4) 918 × 61

5) 1,008 × 13

6) 1,982 × 17

7) 9,802 × 150

8) 10,187 × 22

9) 1,201 × 301

10) 81,500 × 9

Division with 2-digit and 3-digit divisor

Find the quotient:

1) 12 ⟌ 408
2) 39 ⟌ 351
3) 81 ⟌ 891
4) 27 ⟌ 405
5) 99 ⟌ 5,544

6) 102 ⟌ 6,324
7) 905 ⟌ 7,240
8) 175 ⟌ 9,975
9) 184 ⟌ 27,232
10) 821 ⟌ 494,242

Division with remainder

✎ Find the quotient with its remainder:

1) 9)521
2) 87)971
3) 23)652
4) 99)1,005
5) 25)9,807

6) 647)2,009
7) 491)9,000
8) 340)8,745
9) 71)65,109
10) 807)23,659

Estimating quotient

✎ Find the approximate answer:

1) 56 ÷ 15 =
2) 102 ÷ 49 =
3) 385 ÷ 14 =
4) 7,003 ÷ 108 =
5) 2,306 ÷ 30 =

6) 18,420 ÷ 36 =
7) 40,195 ÷ 203 =
8) 16,322 ÷ 799 =
9) 26,981 ÷ 910 =
10) 45,200 ÷ 1,622 =

Multiplying and division numbers ending in zeros

✎ Find the answer:

1) 20 × 800 =
2) 300 × 60 =
3) 9,000 × 400 =
4) 250,000 × 300 =
5) 800,000 × 7,000 =

6) 4,200 ÷ 60 =
7) 80,000 ÷ 2,000 =
8) 3,200,000 ÷ 100 =
9) 60,000,000 ÷ 4,000 =
10) 86,200,000 ÷ 20,000 =

Rounding technique

✎ Round each number to the underlined place value:

1) 9_8_ =
2) 1_2_9 =
3) _4_5,023 =
4) 6_1_,002 =
5) 7_1_,999 =

6) 1_0_,762 =
7) 3,9_4_7,009 =
8) 630,1_7_5 =
9) 39_9_,995 =
10) 800,_0_50 =

Estimation technique

✎ Estimate the following expressions by rounding each number to the nearest hundred (choose a method of your choice):

1) 4,251 + 890 =
2) 2,310 + 9,733 =
3) 7,954 − 1,009 =
4) 61,723 − 14,150 =
5) 650 × 55 =

6) 8,012 × 323 =
7) 28,936 × 9,223 =
8) 14,099 ÷ 625 =
9) 218,625 ÷ 2,514 =
10) 548,712 ÷ 10,999 =

www.mathnotion.com

Word problem

Solve the following problems based on what has been taught so far:

1) Emily received $750 for her birthday. She spent $123.45 on a new skateboard and $219.89 on clothes. Later, she earned $68.50 from doing chores. How much money does Emily have now?

2) A farmer has a total of 7,500 apples. He decides to sell them at the market. Each basket he uses can hold 125 apples, and he plans to take 3 trips to the market.
    a) How many baskets does he need to hold all the apples?
    b) If he sells each basket of apples for $45, how much money will he make per trip?

3) A local library is ordering new books. They plan to buy 38 different titles, and they want 24 copies of each title. Additionally, they are purchasing 56 sets of encyclopedias, and each set contains 9 volumes. How many individual books and volumes will the library have in total once they receive their order?

4) The school cafeteria prepared 12,480 cookies for the school festival. They want to distribute these cookies evenly among 320 students and teachers. Additionally, each class has a cookie jar that needs to be filled with 15 cookies each, and there are 24 classes. How many cookies will each student and teacher get after filling the cookie jars?

5) A construction company is planning to build a new residential area with 52 houses. Each house is estimated to need around 18,500 bricks for construction, but due to potential miscalculations, each house might require between 17,500 to 19,500 bricks. Estimate the total number of bricks needed for the entire project if they plan for the average brick usage per house. Additionally, estimate the maximum and minimum number of bricks needed based on the possible variation in brick requirements per house.

Answer of Worksheets

Adding and subtracting whole number

1) 12,001
2) 15,003
3) 18,614
4) 1,897
5) 25,101
6) 3,883
7) 401
8) 1,909
9) 6,877
10) 7,016

Multiplying whole numbers

1) 2,465
2) 927
3) 22,253
4) 55,998
5) 13,104
6) 33,694
7) 1,470,300
8) 224,114
9) 361,501
10) 733,500

Division with 2-digit and 3-digit divisor

1) 34
2) 9
3) 11
4) 15
5) 56
6) 62
7) 8
8) 57
9) 148
10) 602

Division with remainder

1) 57, R: 8
2) 11, R: 14
3) 28, R: 8
4) 10, R: 15
5) 392, R: 7
6) 3, R: 68
7) 18, R: 162
8) 25, R: 245
9) 917, R:2
10) 29, R: 256

Estimating quotient

1) 4
2) 2
3) 28
4) 70
5) 80
6) 500
7) 200
8) 20
9) 30
10) 30

Multiplying and division numbers ending in zeros

1) 16,000
2) 18,000
3) 3,600,000
4) 75,000,000
5) 5,600,000,000
6) 700
7) 40
8) 32,000
9) 15,000
10) 4,310

www.mathnotion.com

Rounding technique (whole number)

1) 100
2) 100
3) 50,000
4) 61,000
5) 72,000
6) 11,000
7) 3,950,000
8) 630,180
9) 400,000
10) 800,100

Estimation technique (whole number)

1) 5,100
2) 12300
3) 7,000
4) 47,500
5) 42,000
6) 2,400,000
7) 265,880,000
8) 24
9) 87
10) 50

Word problem

1) $475.16
2) a: 60, b: $900
3) 1,416 individual books and volume.
4) 37 cookies.
5) Average estimate: 962,000 bricks; Maximum estimate: 1,014,000 bricks; Minimum estimate: 910,000 bricks.

Chapter 2: Integers

Topics that you will learn in this chapter:

- ➢ Understanding Integers and Opposite Integers
- ➢ Integers and Absolute Value
- ➢ Ordering Integers and Numbers
- ➢ Adding and Dividing Integers
- ➢ Multiplying and Dividing Integers
- ➢ Order of Operations
- ➢ Word problems
- ➢ Worksheets

Integers and Opposite Integers

Integers: Integers include all whole numbers and their negatives $(-3, -2, -1, 0, 1, 2, 3 \ldots)$.

- Positive numbers $(1, 2, 3, \ldots)$
- Negative numbers $(-1, -2, -3, \ldots)$
- Zero (0)

Number Line: Visualize a number line with zero in the center. Positive integers are to the right of zero, and negative integers are to the left:

Opposite Integers: Opposite integers are pairs of numbers that are equal distance from zero but on opposite sides of the number line.

- The opposite of a positive number is a negative number.
- The opposite of a negative number is a positive number.
- The opposite of 0 is 0

Properties:

- Sum to zero: Adding opposite integers always equals zero $(-5 + 5 = 0)$.
- Reflective: Opposites reflect each other across zero on the number line.

Examples

1) Find the number which is omitted.

a)

b)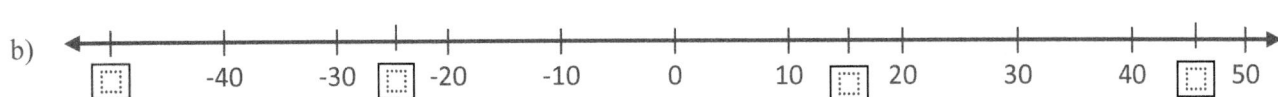

Solution:
   a) From left to right: -2, 4.
   b) From left to right: -50, -25, 15, 45.

2) What is the opposite of the numbers: $-18, 9, -6, 0, 14$

Solution:

From left to right: $18, -9, -6, 0, -14$.

Integers and Absolute Value

Absolute value is the distance a number is from zero on the number line, regardless of direction. It is always a non-negative number.

- Symbol: $|x|$, where x is any number.
- Example: The absolute value of -5 is written as $|-5|$ and equals 5.

Steps to Find Absolute Value:

- Identify the number you're finding the absolute value for.
- Determine how far that number is from zero on the number line.
- The absolute value is that distance, always positive.

Examples:

1) What is the absolute value of -15?

   Solution:

   1. Identify the number: -15.
   2. Find the distance from 0: 15 units.

2) Compare the absolute values of -7 and 3. Which one is larger?

   Solution:

   1. Absolute value of -7: $|-7| = 7$
   2. Absolute value of 3: $|3| = 3$
   3. Compare the values: $7 > 3$

   So, the absolute value of -7 is larger than the absolute value of 3.

3) If the absolute value of a number is 9, what are the possible values for the original number?

   Solution:
   1. Define the equation using absolute value: $|x| = 9$
   2. Find the possible values: $x = 9$ or $x = -9$

   So, the possible values for the original number are 9 and -9.

4) Choose the right sign (> = <): $|-7| - |5| \boxed{\phantom{<}} - 2 + |-6|$

   Solution:

   $|-7| = 7, |5| = 5, |-6| = 6$

   so $|-7| - |5| = 2$ and $-2 + |-6| = 4$

   and finally, $|-7| - |5| \boxed{<} -2 + |-6|$.

www.mathnotion.com

Ordering Integers and Numbers

Ordering integers and numbers is about arranging them from least to greatest (ascending) or greatest to least (descending). Here's how to do it:

1. **List the Numbers.**
2. **Compare each Number**: Identify the smallest and largest.

   On a number line, the integer farthest to the right is always the greatest.

   The integer farthest to the left is always the smallest.

   For example, comparing -2 and -5:

   -2 is to the right of -5, so -2 is greater than -5.

3. **Arrange in Order:**
   - Identify the negative numbers and arrange them in order (the most negative is the smallest).

   Write zero if it's on the list.

   Then write the positive numbers in order.

   - Ascending: From the smallest to the largest.
   - Descending: From the largest to the smallest.

Examples:

✱ Order each set of the following integers.
   a) $5, -2, -15, 8, -8$ (from the smallest to the largest)
   b) $0, -10, 14, 6, -27$ (from the largest to the smallest)

Solution:

a) List the numbers and then identify the smallest and largest:
   - Smallest: -15
   - Largest: 8

   Compare each number: compare -15 with -8: $-15 < -8$. Compare -8 with -2: $-8 < -2$. Compare -2 with 5: $-2 < 5$. Compare 5 with 8: $5 < 8$.

   Arrange in order: $-15, -8, -2, 5, 8$.

b) List the numbers and then identify the smallest and largest:
   - Largest: 14
   - Smallest: $, -27$

   Compare each number: compare 14 with 6: $14 > 6$. Compare 6 with 0: $6 > 0$. Compare 0 with -10: $0 > -10$. Compare -10 with -27: $-10 > -27$.

   Arrange in order: $14, 6, 0, -10, -27$.

www.mathnotion.com

Adding and Subtracting Integers

For adding integer numbers, if the numbers have the same sign, we must add the absolute values and keep the sign, and if the numbers have different signs, it is enough to subtract the smaller absolute value from the larger absolute value and keep the sign of the larger absolute value.

For subtracting integers, we should change the subtraction operation to addition and change the sign of the second number.

Examples

1) Find the result of following expression:
   a) $9 + (-6) =$ b) $(-7) + 6 + (-8) =$

   Solution:
   a) Start with the first number (9) and then add -6, when adding a negative number, it's like subtracting by keeping the sign of the 9. So, the result of $9 + (-6)$ is 3.

   b) First start with the first number (-7), then add 6, since -7 and 6 have different, subtract 6 from 7 and keep the sign of -7 this gives us: $-7 + 6 = -1$. Finally, add -8 (-1 and -8 have the same sign and we must add 1 and 8 and then keep the negative sign), so $-8 - 1 = -9$.

2) Find the result of the following subtraction:
   a) $5 - 7 =$ b) $1 - 4 - 3 =$

   Solution:
   a) We change the $5 - 7$ to $5 + (-7)$. Now, start with the first number (5), then add -7, since 5 and -7 have different signs, we must subtract the 5 from the 7 and keep the sign of -7, therefore: $5 + (-7) = -2$.

   b) We change the $1 - 4 - 3$ to $1 + (-4) + (-3)$. Then start with the number 1 and add -4, since 1 and -4 have different signs, we must subtract the 1 from the 4 and keep the sign of -4, so: $1 + (-4) = -3$. Next add -3 to -3, since both numbers have the same sign, the result is -6.

3) Choose the right sign for following expressions ($> = <$):
   a) $1 - 8 \square 8 - 1$ b) $-9 + (-3) \square -9 - 3$ c) $2 + (-8) \square -3 + 10$

   Solution:

   According to the rules mentioned above:

   a) $1 - 8 = -7$ and $8 - 1 = 7$ so, $1 - 8 \boxed{<} 8 - 1$;

   b) $-9 + (-3) = -12$ and $-9 - 3 = -12$ so, $-9 + (-3) \boxed{=} -9 - 3$;

   c) $2 + (-8) = -6$ and $-3 + 10 = 7$ so, $2 + (-8) \boxed{<} -3 + 10$;

www.mathnotion.com

Multiplying and Dividing Integers

Multiplying and dividing integers follow some straightforward rules. There are two simple steps for each operation:

Multiplying integers:

a) **Multiply the Absolute Values:** Ignore the signs and multiply the numbers as if, they were positive.

b) **Determine the Sign of Product:**
- If both integers have the same sign (both positive or both negative), the product is positive.
- If the integers have different signs (one positive and one negative), the product is negative.

Dividing integers:

a) **Divide the absolute values:** Ignore the signs and divide the numbers as if, they were positive.

b) **Determine the sign of the quotient:**
- If both integers have the same sign (both positive or both negative), the quotient is positive.
- If the integers have different signs (one positive and one negative), the quotient is negative.

Remember: Same Signs = Positive, Different Signs = Negative

Examples:

1) Find the answer of $(-5) \times 7$;

   Solution:

   Step 1: Multiply the absolute values: $5 \times 7 = 35$;

   Step 2: Determine the sign: -5 and 7 have different signs, so the product is negative: $(-5) \times 7 = -35$;

2) Find the answer of $16 \div (-8)$;

   Solution:

   Step 1: Divide the absolute values: $16 \div 8 = 2$;

   Step 2: Determine the sign:

   16 and -8 have different signs, so the quotient is negative: $16 \div (-8) = -2$.

www.mathnotion.com

Order of Operations

The order of operations in math is a set of rules that ensure calculations are performed in the correct sequence, avoiding confusion and errors. If we don't agree on a method to solve math expressions, we could get different answers for the same problem. With the **Order of Operations**, **everyone** should end up with the **same answer**. In general, it's summarized by the acronym **PEMDAS**:

a) **P**arentheses: Do calculations inside parentheses first.
   - ☑ If there is more than one parenthesis, start with the innermost parentheses.
b) **E**xponents: Next, handle exponents (power and roots which are discussed in future chapters).
c) **M**ultiplication and **D**ivision: From left to right.
d) **A**ddition and **S**ubtraction: From left to right.

Common Mistakes to Avoid:

- ✓ Not following left-to-right rule for multiplication/division or addition/subtraction.
- ✓ Skipping parentheses first.
- ✓ Mixing up exponents and multiplication.

Example:

✱ Solve the following expressions:
   a) $7 + (2 \times (-3)) - 8 \div 4$:
   b) $-32 \times (28 - (36 \div 9)) + 120$:

Solution:

a) $7 + (2 \times (-3)) - 8 \div 4$:

   Step 1: Parentheses: $2 \times (-3) = -6$;
   Step 2: Multiplication/Division: $8 \div 4 = 2$;
   Step 3: Addition/Subtraction: $7 + (-6) - 2 = -1$.

b) $-32 \times (28 - (36 \div 9)) + 120$:

   Step 1: Parentheses: Start with the innermost parentheses: $36 \div 9 = 4$;
   Now the expression is: $-32 \times (28 - 4) + 120$;
   Step 2: Parentheses: simplify inside the parentheses: $28 - 4 = 24$;
   Step 3: Multiplication: multiply two numbers 24 and -32 : $-32 \times 24 = -768$;
   Step 4: Addition: add the remaining numbers: $-768 + 120 = -648$.

www.mathnotion.com

Word Problems

To solve word problems involving integer numbers, follow the general process outlines in the first chapter (Reading and Understanding, Translating, Performing the Operations, Simplifying and Checking Work). additionally, it is crucial to identify the type of integer operations involved adhere to the order of operations (PEMDAS) when solving the expression or equation.

Example

Jamie is exploring a cave that descends below sea level. He starts at the entrance, which is at sea level (0 feet). He climbs down 25 feet to the first chamber, then further down another 30 feet to the second chamber. After exploring the second chamber, he climbs back up 15 feet. Finally, he descends another 40 feet to reach the lowest chamber.

How deep is Jamie now compared to sea level?

Solution:

Step 1: Understand the Problem:

- Jamie starts at sea level (0 feet).
- He climbs down 25 feet.
- Climb further down another 30 feet.
- Climbs back up 15 feet.
- Descends another 40 feet

Step 2: Highlight important information:

- Starting point: 0 feet
- Descends: 25 feet, 30 feet, and 40 feet
- Climbs up: 15 feet.

Step 3: Translate to math:

- Starting at 0 feet: 0
- Descend 25 feet: 0−25.
- Further down 30 feet: −25−30
- Climb up 15 feet: −55+15.
- Descend another 40 feet: −40−40

Step 4: perform the operations:

- Start: 0
- Descend 25 feet: 0−25=−25
- Further down 30 feet: −25−30=−55
- Climb up 15 feet: −55+15=−40
- Descend another 40 feet: −40−40=−80

Final Answer: Jamie is 80 feet below sea level.

www.mathnotion.com

Worksheets

Understanding integers and opposite integers

✎ Write the opposite of each number:

1) -7 2) 15 3) – (-20) 4) 161 5) - 33 6) – (-(-8)) 7) 12

✎ Write the opposite of each number relative to the given number on its right side:

8) 5 (relative to −1) 9) -3 (relative to 2) 10) 11 (relative to the −5)

Integers and Absolute Value

✎ Write absolute value of each number:

1) $|-2| =$
2) $|-27| =$
3) $|-20| =$
4) $|-(-14)| =$
5) $|-6| =$

✎ Evaluate the value:

6) $|-5 + 8 - 9| =$
7) $|-2| - |-8 + 2| =$
8) $|-20 \div 5| \times 2 =$
9) $|-16 \div (-8) - |-4|| - |-10| =$
10) $|-(-12) + |-9 \times 3| \div (-9)| =$

Ordering Integers and Numbers

✎ Order each set of integers from least to greatest:

1) $8, -10, -5, -3, 4$ ___, ___, ___, ___, ___, ___
2) $-10, -18, 6, 14, 27$ ___, ___, ___, ___, ___, ___
3) $15, -8, -21, 21, -23$ ___, ___, ___, ___, ___, ___
4) $-14, -40, 23, -12, 47$ ___, ___, ___, ___, ___, ___
5) $59, -54, 32, -57, 36$ ___, ___, ___, ___, ___, ___

✎ Order each set of integers from greatest to least:

6) $18, 36, -16, -18, -10$ ___, ___, ___, ___, ___, ___
7) $27, 34, -12, -24, 94$ ___, ___, ___, ___, ___, ___
8) $50, -21, -13, 42, -2$ ___, ___, ___, ___, ___, ___
9) $37, 46, -20, -16, 86$ ___, ___, ___, ___, ___, ___
10) $-18, 88, -26, -59, 75$ ___, ___, ___, ___, ___, ___

www.mathnotion.com

Adding and subtracting integers

✏️ Do the following addition and subtraction operations:

1) $(-7) + (-8) =$
2) $5 + (-12) =$
3) $6 + (-11) + 3 =$
4) $(-9) + 15 + (-4) =$
5) $(-3) + (-10 - 2) + 19 =$

6) $(-11) - (-6) =$
7) $(-14) - 4 =$
8) $8 - (-16) - 1 =$
9) $(12 + 9) - (-8 - 9) =$
10) $-7 - (-12) - (17 - 2) =$

Multiplying and Dividing Integers

✏️ Find each product and quotient:

1) $(-3) \times (-8) =$
2) $(-7) \times 3 \times 6 =$
3) $5 \times (-10) \times 4 =$
4) $(-7 + 5) \times 6 =$
5) $(-3 - 6) \times (-6 + 4 - 2) =$

6) $-42 \div (7) =$
7) $(-30) \div (-6) =$
8) $64 \div (-7 - 1) =$
9) $(-135 \div 5) \div (-9) =$
10) $36 \div (-4 \times 3) =$

Order of Operations

✏️ Evaluate each expression:

1) $14 - (3 \times 6) =$
2) $(19 \times 4) + 16 =$
3) $(16 - 7) - (8 \times 2) =$
4) $27 + (18 \div 3) =$
5) $(18 \times 8) \div 6 =$
6) $(13 + 5 - 14) \times 3 - 2 =$

7) $(85 - 20) + (20 - 18 + 7) =$
8) $32 + (28 - (36 \div 9)) =$
9) $-25 - (60 \div (-12 \times 5))$
10) $-(-(-20 + 8) \times 4 \div 2) =$

Word problems

✏️ Solve the following problems based on what has been taught so far:

1) A submarine is at a depth of 150 meters below sea level. It ascends 40 meters, then descends 25 meters. What is the submarine's final depth relative to sea level?

2) In a game, Alex gained 25 points, lost 40 points, gained another 30 points, and then lost 15 points. What is his final score?

3) At dawn, the temperature was -5°C. By noon, it rose by 12°C. In the afternoon, it dropped by 8°C, and by nightfall, it dropped another 10°C. What was the final temperature at night?

4) A mountain climber starts at an elevation of 1,500 meters above sea level. She ascends 300 meters to reach a new height, then descends 600 meters to set up camp. The next day, she ascends 200 meters and finally descends 900 meters to base camp. What is her elevation at the base camp?

www.mathnotion.com

5) In a city, the temperature at midnight was -10°C. During the early morning, it decreased by another 8°C. By noon, the sun had come out, increasing the temperature by 18°C. By 4 PM, the temperature dropped again by 6°C. What was the temperature at 4 PM?

6) A hiker starts at a cabin 200 meters below sea level. He ascends 250 meters to a trail, then descends 100 meters to a valley, and ascends another 350 meters to a peak. Finally, he descends 500 meters to the campsite. What is the hiker's final elevation relative to sea level?

7) The temperature in a mountain range can be quite variable. At dawn, it was 2°C. It fell by 7°C by mid-morning, then rose by 10°C at noon due to the sun. By evening, the temperature had dropped by 12°C again. What was the temperature by evening?

8) If we consider 12 noon as the starting point and represent the hours before noon with negative integers and the hours after noon with positive integers, express the following hours as integers.
   a) 1 hour before noon
   b) 4:00' am
   c) 15:00'
   d) 8:00' Pm

Answer of Worksheets

Understanding integers and opposite integers

1) 7
2) -15
3) -20
4) -161
5) 33
6) 8
7) -12
8) -7
9) 7
10) -21

Integers and Absolute Value

1) 2
2) 27
3) 20
4) 14
5) 6
6) 6
7) -4
8) 8
9) -8
10) 9

Ordering Integers and Numbers

1) $-10, -5, -3, 4, 8$
2) $-18, -10, 6, 14, 27$
3) $-23, -21, -8, 15, 21$
4) $-40, -14, -12, 23, 47$
5) $-57, -54, 32, 36$
6) $36, 18, , -10, -16, -18$
7) $94, 34, 27, -12, -24$
8) $50, 42, -2, -13, -21$
9) $86, 46, 37, -16, -20$
10) $88, 75, -18, -26, -59$

Adding and subtracting integers

1) -15
2) -7
3) -2
4) 2
5) 4
6) -5
7) -18
8) 23
9) 38
10) -10

Multiplying and Dividing Integers

1) 24
2) -126
3) -200
4) -12
5) 36
6) -6
7) 5
8) -8
9) 3
10) -3

Order of Operations

1) -4
2) 92
3) -7
4) 33
5) 24
6) 10
7) 74
8) 56
9) -24
10) -24

Word problems

1) -135 meters.
2) 0 points.
3) $-11°C$.
4) 500 meters.
5) $-6°C$.
6) -200 meters.
7) $-7°C$.
8) a) -1, b) -8, c) 3, d) 8

www.mathnotion.com

Chapter 3: Number Theory

Topics that you will learn in this chapter:
- Prime and Composite Numbers
- Divisibility Rules
- Prime Factorization
- Greatest Common Factor
- Least Common Multiple
- Word Problems
- Worksheets

Prime and Composite Numbers

Prime Numbers:

A prime number is a natural number greater than 1 that has no positive divisors other than 1 and itself, such as 2, 3, 5, 7, 11, 13, 17, etc.

- ☑ Number 2 is the only prime even number because all other even numbers are divisible by 2.

Composite Numbers:

A composite number is a natural number greater than 1 that has more than two positive divisors such as 4, 6, 8, 9, 10, 12, 14, etc.

- ☑ Composite numbers can be factored into smaller positive integers (other than 1 and itself).
- ☑ The number 1 is neither prime nor composite.

Steps to Recognize Prime Numbers:

Greater than 1: Ensure the number is greater than 1.

Check Divisibility: Determine if the number has any divisors other than 1 and itself. Start by checking divisibility by 2, 3, 5, 7, etc., up to the square root of the number.

Steps to Recognize Composite Numbers:

Greater than 1: Ensure the number is greater than 1.

Check Divisibility: If the number has any divisors other than 1 and itself, it's composite.

Example:

Check if the numbers 29 and 24 are prime or not:

Solution:

Number: 29

Greater than 1: Yes.

Check divisibility by 2, 3, 5:

Not divisible by 2 (not even).

Not divisible by 3 (sum of digits, 2 + 9 = 11, not divisible by 3 (refer to divisibility rules).

Not divisible by 5 (does not end in 0 or 5(refer to divisibility rules).

29 is prime.

Number: 24

Greater than 1: Yes.

Check divisibility by 2, 3, 4, etc.:

Divisible by 2 (even).

24 is composite.

Divisibility Rules

Divisibility rules are simple shortcuts that help you determine whether one number is divisible by another without performing the division. Here are some key divisibility rules:

- **Divisibility by 2**: A number is divisible by 2 if its last digit is even (0, 2, 4, 6, or 8).
- **Divisibility by 3:** A number is divisible by 3 if the sum of its digits is divisible by 3.
- **Divisibility 4:** A number is divisible by 4 if the last two digits form a number that is divisible by 4.
- **Divisibility by 5**: A number is divisible by 5 if its last digit is 0 or 5.
- **Divisibility by 6:** A number is divisible by 6 if it is divisible by both 2 and 3.
- **Divisibility by 7:** A number is divisible by 7 if doubling the last digit and subtracting it from the rest of the number results in a number that is divisible by 7.
- **Divisibility by 8:** A number is divisible by 8 if the last three digits form a number that is divisible by 8.
- **Divisibility by 9:** A number is divisible by 9 if the sum of its digits is divisible by 9.
- **Divisibility by 10:** A number is divisible by 10 if its last digit is 0.

Example:

Check if the following numbers are divisible by the number in parenthesis?

a) 126 (divisible by 3?)

b) 1928 (divisible by 4?)

c) 231 (divisible by 7?)

d) 199 (divisible by 9?)

e) 73 (divisible by 2?)

f) 1056 (divisible by 8?)

g) 62580 (divisible by 5?)

h) 6990 (divisible by 6?)

i) 5003 (divisible by 10?)

Solution:

a) Yes, because the sum of its digits $(1 + 2 + 6 = 9)$ is divisible by 3.

b) Yes, because the last two digits from 1928 (28) are divisible by 4.

c) Yes, because doubling the last digit (2) and subtracting it from the rest of the 231 (23) result in 21, $(23 - 2 = 21)$ which is divisible by 7.

d) No, because the sum of its digits $(1 + 9 + 9 = 19)$ is not divisible by 9.

e) No, because its last digit (3) is odd.

f) Yes, because its last three digits (056) are divisible by 8.

g) Yes, because its last digit (0) is 0.

h) Yes, because its last digit is even so it is divisible by 2, and the sum of its digits $(6 + 9 + 9 + 0 = 21)$ is divisible by 3.

i) No, because its last digit (3) is not 0.

www.mathnotion.com

Identify Factors

Identifying factors of a number involves finding all the whole numbers that can divide that number without leaving a remainder. Here's how you can do it:

Steps to Identify Factors:

1. **Understand Factors:** Factors are numbers that can be multiplied together to get another number. For example, factors of 12 are numbers that multiply together to make 12 (like 3 and 4 because $3 \times 4 = 12$).
2. **Start with 1:** Always start with 1 and the number itself. These are always factors.
3. **Test Small Numbers**: Test other small numbers to see if they divide evenly into the number. If they do, they are factors.
   - ☑ You should note that this process can be continued until reaching a repeated factor, then the process is finished.
4. **List All Pairs:** Write down pairs of numbers that multiply to the original number.
5. **Compile the List:** Gather all the factors from your pairs.

Examples:

1) Identify the factors of 24:
   Solution:
   1. Start with 1 and 24:
      - 1 and 24 are always factors because $1 \times 24 = 24$.
   2. Smaller Numbers Test:
      - 2: Check if 24 is divisible by 2: Yes, $2 \times 12 = 24$. So, 2 and 12 are factors.
      - 3: Check if 24 is divisible by 3: Yes, $3 \times 8 = 24$. So, 3 and 8 are factors.
      - 4: Check if 24 is divisible by 4: yes, $4 \times 6 = 24$. So, 4 and 6 are factors.
      - 5: Check if 24 is divisible by 5: No, its last digit is not 0 or 5.
      - 6: Check if 24 is divisible by 6: Yes, but 6 is repetitive! So, we stop the process.
   3. List All Pairs: $1 \times 24, 2 \times 12, 3 \times 8, 4 \times 6$;
   4. Compile the List of Factors: $1, 2, 3, 4, 6, 8, 12, 24$.

2) Identify the factors of 15:
   Solution:
   1. Start with 1 and 15:
      - 1 and 15 are always factors because $1 \times 15 = 15$
   2. Smaller Numbers Test:
      - 2: Check if 15 is divisible by 2: No, 15 is odd number.
      - 3: Check if 15 is divisible by 3: Yes, $3 \times 5 = 15$. So, 3 and 5 are factors.
      - 4: Check if 15 is divisible by 4: No, 15 is not divisible by 5.
      - 5: Check if 15 is divisible by 5: Yes, but 5 is repetitive! So, we stop the process.
   3. List All Pairs: $1 \times 15, 3 \times 5$.
   4. Compile the List of Factors: $1, 3, 5, 15$.

www.mathnotion.com

Prime Factorization

Prime factorization is the process of breaking down a composite number into the prime numbers that multiply together to equal the original number. Here's how you can do it:

Steps for Prime Factorization:

1. **Start with the Smallest Prime Number**: Begin with 2, the smallest prime number.
2. **Divide the Number by the Prime**: If the number is divisible by 2, divide it and continue with the quotient. If not, move to the next prime number (3, 5, 7, etc.).
3. **Repeat the Process**: Keep dividing the quotient by prime numbers until you can't divide anymore.
4. **List All Prime Factors**: The prime factors are the numbers you used for division.

Examples:

1) Find the prime factors of 12:

   Solution:
   1. Start with the Smallest Prime Number: Begin with 2, the smallest Prime number.
   2. Divide by 2: $12 \div 2 = 6$. So, 2 is a prime factor, and we continue with quotient 6.
   3. Continue with 2: $6 \div 2 = 3$. 2 is a prime factor again, and we continue with quotient 3.
   4. Move to the Next Prime: The quotient is now 3, which is a prime number itself.

   So, the prime factors of 12 are $2 \times 2 \times 3$.

2) Find the prime factors of 30:

   Solution:
   1. Start with the Smallest Prime Number: Begin with 2, the smallest Prime number.
   2. Divide by 2: $30 \div 2 = 15$. So, 2 is a prime factor, and we continue with quotient 15.
   3. Move to the Next Prime Number (which is 3): $15 \div 3 = 5$. 3 is a prime factor, and we continue with quotient 5.
   4. Stop at Prime Factor 5: 5 is a prime number itself.

   So, the prime factors of 30 are $2 \times 3 \times 5$.

3) What are the number of different prime factors of 120:

   Solution:
   According to the process of finding prime factors of 120 we have: $120 = 2 \times 2 \times 2 \times 3 \times 5$. So, the number of different prime factors of 120 is 3 (2, 3, 5).

www.mathnotion.com

Greatest Common Factor

The Greatest Common Factor (GCF) of two or more numbers is the largest number that divides all of them without leaving a remainder. Here's how to find it:

Steps to Find the GCF:

1. **List the Factors**: Write down all factors of each number.
2. **Identify Common Factors:** Find the numbers that are common to all lists.
3. **Choose the Greatest:** The largest common number is the GCF.
☑ **Another Method: Prime Factorization**
   - Find prime factorization of each number.
   - Identify common prime factors.
   - Multiply common prime factors.

Examples:

1) Find the GCF of 24 and 36:
   Solution:
   1. List the Factors:
      - Factors of 24: 1, 2, 3, 4, 6, 8, 12, 24
      - Factors of 36: 1, 2, 3, 4, 6, 9, 12, 18, 36
   2. Identify Common Factors:
      - Common factors: 1, 2, 3, 4, 6, 12
   3. Choose the Greatest: The greatest common factor is 12.

2) Find the GCF of 72 and 48 (use the Prime Factorization Method):
   Solution:
   1. Prime Factorization:
      - Prime factors of 72: $72 = 2 \times 2 \times 2 \times 3 \times 3$
      - Prime factors of 48: $48 = 2 \times 2 \times 2 \times 2 \times 3$
   2. Identify Common Prime Factors:
      - Common prime factor: $2 \times 2 \times 2 \times 3$
   3. Multiply the Common Prime Factors: GCF= 24.

3) Find the GCF of 12, 18, 6 (use the Prime Factorization Method):
   Solution:
   According to the steps you have seen in 2 past examples:
   Prime factors of $12 = 2 \times 2 \times 3$
   Prime factor of $18 = 2 \times 3 \times 3$
   Prime factor of $6 = 2 \times 3$
   Common prime factor = 2×3. So, the multiplying of common prime factors result in 6.

www.mathnotion.com

Least Common Multiple

The Least Common Multiple (LCM) of two or more numbers is the smallest number that is a multiple of all the numbers. It's useful in various mathematical problems, especially when adding, subtracting, or comparing fractions with different denominators.

Steps to Find the LCM:

1. **List the Multiples**: Write down a list of multiples for each number.
2. **Find the Common Multiples**: Identify the common multiples in each list.
3. **Choose the Smallest Multiple**: The smallest multiple common to all lists is the LCM.
☑ **Another Method: Prime Factorization:**
   - Find the prime factorization of each number.
   - Use the highest repetition of each prime and the primes which are unique in each number.

Examples:

1) Find the LCM of 12 and 18.
   Solution:
   1. List the Multiples:
      - Multiples of 12: 12, 24, 36, 48, 60, 72, 84, 96 ...
      - Multiples of 18: 18, 36, 54, 72, 90 ...
   2. Find the Common Multiples:
      - Common multiples: 36, 72...
   3. Choose the Smallest Multiple:
      - The smallest common multiple: 36.

2) Find the LCM of 15, 10, 12.
   Solution:
   1. List the Multiples:
      - Multiples of 10: 10, 20, 30, 40, 50, 60, 70, 80, 90, 100, 110, 120
      - Multiples of 12: 12, 24, 36, 48, 60, 72, 84, 96, 108, 120 ...
      - Multiples of 15: 15, 30, 45, 60, 75, 90, 105, 120 ...
   2. Find the Common Multiples:
      - Common multiples: 60, 120 ...
   3. Choose the Smallest Multiple:
      - The smallest common multiple: 60.

3) Find the LCM of 49 and 21 by using Prime Factorization method:
   Solution:
   Prime factors of 49: $49 = 7 \times 7$
   Prime factors of 21: $21 = 3 \times 7$
   So, LCM = $7 \times 7 \times 3$.

www.mathnotion.com

Word Problems

Solving word problems about number theory involves a clear understanding of various concepts and applying them systematically. The original steps are the same as the previous chapters (Reading and Understanding, Translating, Performing the Operations, Simplifying and Checking Work), and applying the relevant method according to the topics which were discussed in this chapter, is crucial.

Examples:

1) The number of students in a class is between 24 and 30. If the class teacher cannot divide the students into equal groups, how many students can there be in the class?

   Solution:
   1. **Understand the Problem:** We need to find a number between 24 and 30 that cannot be divided into equal groups (except for groups of 1 and the number itself).
   2. **Identify Possible Numbers:** Numbers between 24 and 30: 25, 26, 27, 28, 29.
   3. **Check for Prime Numbers:** Prime numbers are only divisible by 1 and themselves. Let's find the prime numbers between 24 and 30.
      - 25: Not prime (divisible by 5 and 5).
      - 26: Not prime (divisible by 2 and 13).
      - 27: Not prime (divisible by 3 and 9).
      - 28: Not prime (divisible by 2, 4, 7, 14).
      - 29: Prime (only divisible by 1 and 29).
   4. **Conclusion:** The class can have 29 students because it cannot be divided into equal groups.

2) You have two pieces of rope, one 18 meters long and the other 24 meters long. You want to cut them into pieces of equal length without any leftover rope. What is the greatest length each piece can be?

   Solution:
   1. Identify the Problem: This is a GCF problem.
   2. Prime Factorization:
      - 18: $2 \times 3 \times 3$
      - 24: $2 \times 2 \times 2 \times 3$
   3. Find the GCF:
      - Common factors: $2 \times 3 = 6$.

Worksheets

Prime and composite numbers

✍ Check if the following numbers are prime or composite:

1) 28
2) 13
3) 51
4) 44
5) 97
6) 79
7) 231
8) 67
9) 188
10) 171

Divisibility rules

✍ Check if the following numbers are divisible by the number in parenthesis?

1) 26 (divisible by 2?)
2) 269 (divisible by 9?)
3) 1,008 (divisible by 8?)
4) 350 (divisible by 5?)
5) 3,331 (divisible by 3?)
6) 210 (divisible by 10?)
7) 161 (divisible by 7?)
8) 423 (divisible by 4?)
9) 900 (divisible by 6?)
10) 1,254 (divisible by 3?)

Identify factors.

✍ List all positive factors of each number:

1) 16
2) 28
3) 95
4) 56
5) 63
6) 80
7) 72
8) 100
9) 65
10) 91

Prime factorization

✍ List the prime factorization for each number:

1) 10
2) 26
3) 30
4) 44
5) 78
6) 105
7) 96
8) 40
9) 68
10) 39

Greatest common factor

✍ Find the GCF for each number pair:

1) 4, 36
2) 6, 10
3) 7, 3
6) 9, 2, 3
7) 5, 15, 10
8) 6, 8, 12

6th Grade Nevada Math

4) 42, 14
5) 28, 52

9) 100, 101
10) 180, 120

Least common multiple

🖉 Find the LCM for each number pair:

1) 6, 9
2) 15, 45
3) 8, 20
4) 22, 11
5) 4, 11

6) 30, 54
7) 2, 3, 4
8) 10, 9
9) 12, 15, 60
10) 24, 72, 96

Word problem

🖉 Do the following word problems according to what you learned in this chapter:

1) Jack and Jill are training for a marathon. Jack runs a loop every 8 minutes, while Jill runs a loop every 12 minutes. If they both start running at the same time, after how many minutes will they be running together again at the starting point?

2) Carlos wants to organize a party and needs to create goody bags for his guests. He has 144 candies and 108 chocolates. He wants each goody bag to have the same number of candies and chocolates without any leftovers. How many goody bags can Carlos make, and how many candies and chocolates will each goody bag have?

3) Pencils come in packages of 10. Erasers come in packages of 12. Phillip wants to purchase the smallest number of pencils and erasers so that he will have exactly 1 eraser per pencil. How many packages of pencils and erasers should Phillip buy?

4) Beginning at 8:30 A.M., tours of the National Capitol and the White House begin at a tour agency. Tours for the National Capitol leave every 15 minutes. Tours for the White House leave every 20 minutes. How often do the tours leave at the same time?

5) There are 48 girls and 64 boys in the school choir. The choir teacher plans to arrange the students in equal rows. Only girls or boys will be in each row. What is the greatest number of students that could be in each row?

www.mathnotion.com

Answer of Worksheets

Prime and composite numbers
1) Composite
2) Prime
3) Composite
4) Composite
5) Prime
6) Prime
7) Composite
8) prime
9) Composite
10) Composite

Divisibility Rules
1) Yes
2) No
3) Yes
4) Yes
5) No
6) Yes
7) Yes
8) No
9) Yes
10) Yes

Identify factors.
1) 1, 2, 4, 8, 16
2) 1, 2, 4, 7, 14, 28
3) 1, 5, 19, 95
4) 1, 2, 4, 7, 8, 14, 28, 56
5) 1, 3, 7, 9, 21, 63
6) 1, 2, 4, 5, 8, 10, 16, 20, 40, 80
7) 1, 2, 3, 4, 6, 8, 9, 12, 18, 24, 36, 72
8) 1, 2, 4, 5, 10, 20, 25, 50, 100
9) 1, 5, 13, 65
10) 1, 7, 13, 91

Prime factorization
1) 2×5
2) 2×13
3) $2 \times 3 \times 5$
4) $2 \times 2 \times 11$
5) $2 \times 3 \times 13$
6) $3 \times 5 \times 7$
7) $2 \times 2 \times 2 \times 2 \times 2 \times 3$
8) $2 \times 2 \times 2 \times 5$
9) $2 \times 2 \times 17$
10) 3×13

Greatest common factor
6) 4
7) 2
8) 1
9) 14
10) 4
6) 1
7) 5
8) 2
9) 1
10) 60

www.mathnotion.com

Least common multiple

1) 18
2) 45
3) 40
4) 22
5) 44
6) 270
7) 12
8) 90
9) 60
10) 288

Word problem

1) 24 minutes.
2) 36 goody bags, each containing 4 candies and 3 chocolates.
3) 6 packages of pencils and 5 packages of erasers.
4) Every 60 minutes.
5) Equal rows of 16 students.

Chapter 4: Fractions

Topics that you will learn in this chapter:
- Converting Between Fractions, Mixed Numbers
- Ordering Fractions
- Simplifying Fractions
- Least Common Denominator
- Adding and Subtracting (Fractions and Mixed Numbers)
- Multiplying and Dividing (Fractions and Mixed Numbers)
- Mixed Operations
- Word Problems
- Worksheets

Fractions and Mixed Numbers

When the numerator of a fraction is greater than the denominator, the fraction is greater than one (we call it improper fractions), and it can be converted into a mixed number. Here's how you can do it:

Converting Improper Fractions to Mixed Numbers:

1. **Divide the Numerator by the Denominator**: This gives you the whole number part.
2. **Find the Remainder**: This is the numerator of the fractional part.
3. **Write the Mixed Number**: Combine the whole number and the fraction.

Converting Mixed Numbers to Improper Fractions:

1. **Multiply the Whole Number by the Denominator**: This gives you part of the new numerator.
2. **Add the Numerator**: This completes the new numerator.
3. **Write the Improper Fraction:** Combine the new numerator and the original denominator.

Examples:

1) Convert $\frac{13}{5}$ to a mixed number:

   Solution:
   1. Divide 13 by 5: The result is 2 with a remainder of 3.
   2. The mixed number is $2\frac{3}{5}$.

$$5\overline{)13} \atop \underline{-10} \atop 3$$ (quotient 2)

2) Convert $3\frac{5}{6}$ to an improper fraction:

   Solution:
   1. Multiply 3 (whole number) by 6 (denominator): $3 \times 6 = 18$.
   2. Add the numerator (5): $18 + 5 = 23$.
   3. The improper fraction is $\frac{23}{6}$.

3) Find the missing number.

   a) $\frac{16}{5} = 3\frac{\square}{5}$ b) $2\frac{\square}{7} = \frac{17}{7}$ c) $\square\frac{2}{9} = \frac{38}{9}$ d) $\frac{15}{8} = 1\frac{\square}{\square}$

   Solution: We can convert the improper fraction to mixed numbers to find the missing number.

   $5\overline{)16}$ quotient 3, remainder 1
   $7\overline{)17}$ quotient 2, remainder 3
   $9\overline{)38}$ quotient 4, remainder 2
   $8\overline{)15}$ quotient 1, remainder 7

   a) $\frac{16}{5} = 3\frac{1}{5}$ b) $2\frac{3}{7} = \frac{17}{7}$ c) $4\frac{2}{9} = \frac{38}{9}$ d) $\frac{15}{8} = 1\frac{7}{8}$

Ordering Fractions

There are many ways to order fractions. We must choose the right method according to the given fractions. First of all, it's better to separate proper fractions and improper fractions and then use an appropriate method to order them. Here are some practical ways to do it:

1. **Common denominators:**
   - Convert all fractions to have the same denominator. This makes it easier to compare the numerators directly.
   - Find a common denominator.
   - Compare numerators and then order them.
2. **Convert to decimals:**
   Another method is to convert fractions to decimals by dividing the numerator by the denominator.
3. **Cross-Multiplication:**
   Use cross-multiplication to compare two fractions directly without finding a common denominator.

☑ Sometime converting improper fractions to mixed numbers makes this process more easily.

☑ When all fractions are proper fractions, we can compare them with $\frac{1}{2}$, if the numerator is greater than half of denominator, then the fraction is greater than $\frac{1}{2}$, and if the numerator is less than half of denominator, then the fraction is less than $\frac{1}{2}$.

Examples:

1) Order the following fractions from least to greatest:
   $\frac{10}{3}, \frac{1}{5}, \frac{5}{2}, \frac{4}{6}, \frac{12}{9}, \frac{8}{10}$

   Solution:
   - First convert improper fraction to mixed number: $\frac{10}{3} = 3\frac{1}{3}, \frac{5}{2} = 2\frac{1}{2}, \frac{12}{9} = 1\frac{3}{9}$.
   - Order mixed numbers according to their whole numbers: $1\frac{3}{9} < 2\frac{1}{2} < 3\frac{1}{3}$.
   - Find the common denominators for ordering proper fractions: for $\frac{1}{5}, \frac{4}{6}, \frac{8}{10}$ the GCF is 30, so we have: $\frac{1}{5} = \frac{6}{30}, \frac{4}{6} = \frac{20}{30}, \frac{8}{10} = \frac{24}{30}$.
   - Order proper fraction by comparing their numerators: $\frac{6}{30} < \frac{20}{30} < \frac{24}{30}$.
   - Final order by matching back to original fractions: $\frac{1}{5} < \frac{4}{6} < \frac{8}{10} < \frac{12}{9} < \frac{5}{2} < \frac{10}{3}$.

2) Compare $\frac{23}{50}$ and $\frac{17}{28}$:

   Solution:
   $\frac{23}{50}$ is less than $\frac{1}{2}$, because 23 is smaller than half of 50 (25). In the other hand, $\frac{17}{28}$ is greater than $\frac{1}{2}$, because 17 is greater than half of 28 (14), so we have: $\frac{23}{50} < \frac{1}{2} < \frac{17}{28}$, therefore $\frac{23}{50} < \frac{17}{28}$.

www.mathnotion.com

Simplifying Fractions

Simplifying fractions means reducing them to their simplest form where the numerator and denominator have no common factors other than 1. Here's how you do it:

Steps to Simplify Fractions:

1. **Find the Greatest Common Factor (GCF):** Identify the largest number that can divide both the numerator and the denominator.
2. **Divide Both the Numerator and Denominator by the GCF:** This will give you the fraction in its simplest form.

Example:

1) Simplify the following fractions:

   a) $\frac{72}{64}$ b) $\frac{120}{80}$ c) $\frac{66}{55}$ d) $\frac{35}{75}$

Solution:

a) $\frac{72}{64}$:
   - Find the greatest common factor: $72 = 2 \times 2 \times 2 \times 3 \times 3$ and $64 = 2 \times 2 \times 2 \times 2 \times 2 \times 2$ So, the GCF is $2 \times 2 \times 2 = 8$.
   - Divide both by the GCF: $72 \div 8 = 9$ and $64 \div 8 = 8$.
   - Simplified fraction: $\frac{9}{8}$.

b) $\frac{120}{80}$:
   - Find the greatest common factor: $120 = 2 \times 2 \times 2 \times 3 \times 5$ and $80 = 2 \times 2 \times 2 \times 2 \times 5$ So, the GCF is $2 \times 2 \times 2 \times 5 = 40$.
   - Divide both by the GCF: $120 \div 40 = 3$ and $80 \div 40 = 2$.
   - Simplified fraction: $\frac{3}{2}$.

c) $\frac{55}{66}$:
   - Find the greatest common factor $55 = 5 \times 11$: and $66 = 2 \times 3 \times 11$ So, the GCF is 11.
   - Divide both by the GCF: $55 \div 11 = 5$ and $66 \div 11 = 6$.
   - Simplified fraction: $\frac{5}{6}$.

d) $\frac{35}{75}$:
   - Find the greatest common factor: $35 = 5 \times 7$ and $75 = 5 \times 5 \times 3$ so the GCF is 5.
   - Divide both by the GCF: $35 \div 5 = 7$ and $75 \div 5 = 15$.
   - Simplified fraction: $\frac{7}{15}$.

www.mathnotion.com

Least Common Denominator

The Least Common Denominator (LCD) is the smallest number that can be a common denominator for a set of fractions. Using LCD is incredibly Useful because it simplifies the process of working with fractions. Here's a step-by-step guide to finding it:

Steps to Find the LCD by Using the Multiples of Denominators:

1. **Identify the Denominators:** List the denominators of the fractions you're working with.
2. **List the Multiples**: Write down a list of multiples for each denominator.
3. **Find the Common Multiples**: Identify the common multiples in each list.
4. **Choose the Smallest Multiple**: The smallest multiple common to all lists is the LCD.
☑ **Another Method: Prime Factorization:**
   - Find the prime factorization of each denominator.
   - Use the highest repetition of each prime and the primes which are unique in each denominator.

Examples:

1) Find the LCD of $\frac{7}{8}$, $\frac{5}{12}$ and $\frac{9}{16}$:

   Solution:
   1. Identify the Denominators: 8, 12 and 16
   2. List the Multiples:
      - Multiples of 8: 8, 16, 24, 32, 40, 48, 56, 64, 72…
      - Multiples of 12: 12, 24, 36, 48, 60, 72…
      - Multiples of 16: 16, 32, 48, 64…
   3. Find the Common Multiples:
      - Common multiples: 48, 96, …
   4. Choose the Smallest Multiple:
      - The smallest common multiple: 48

2) Find the LCD of $\frac{11}{135}$ and $\frac{14}{90}$ by using Prime Factorization method:

   Solution:
   1. Identify the Denominators: 135 and 90
   2. Find the Prime Factorization of Denominators:
      - The Prime Factors of 135: $135 = 3 \times 3 \times 3 \times 5$
      - Prime factors of 90: $90 = 2 \times 3 \times 3 \times 5$

   So, LCD = $3 \times 3 \times 3 \times 5 \times 2 = 270$

Adding and Subtracting

We can only add or subtract fractions and mixed numbers if they have the same denominators. So, we'll need to find the lowest common denominator before we add or subtract these fractions and mixed numbers. Once the fractions have the same denominator, we can add or subtract as usual. Here is a guide to help you through the process:

1. **Convert all Mixed numbers to Improper Fractions:** Change each mixed number to an improper fraction.
2. **Find the Least Common Denominator (LCD):** Do the process of finding least common denominator as mentioned in precious page.
3. **Add/Subtract the Numerators:** Add or subtract the fractions.
4. **Convert Back to Mixed Numbers:** If the result is an improper fraction, convert it back to a mixed number.
5. **Simplify:** Simplify the resulting fraction or mixed number if possible.

Examples:

3) Add $1\frac{5}{6}$ and $\frac{7}{8}$:

   Solution:
   - Convert mixed numbers to Improper Fractions: $1\frac{5}{6} = \frac{11}{6}$.
   - Find the LCD: As we thought the process of finding LCM in chapter 3, the LCM or the LCD of 6 and 8 using their prime factors is 24. So, we have $\frac{11}{6} = \frac{44}{24}$ and $\frac{7}{8} = \frac{21}{24}$.
   - Add the Numerators: $\frac{44}{24} + \frac{21}{24} = \frac{65}{24}$.
   - Convert Back to Mixed Numbers: $\frac{65}{24} = 2\frac{17}{24}$.
   - Simplify: The fraction $2\frac{17}{24}$ is already in its simplest form.

4) Subtract $\frac{1}{9}$ and $3\frac{4}{12}$:

   Solution:
   - Convert mixed numbers to Improper Fractions: $3\frac{4}{12} = \frac{40}{12}$.
   - Find the LCD: As we thought the process of finding LCM in chapter 3, the LCM or the LCD of 9 and 12 using their prime factors is 36. So, we have $\frac{40}{12} = \frac{120}{36}$ and $\frac{1}{9} = \frac{4}{36}$.
   - Subtract the Numerators: $\frac{120}{36} - \frac{4}{36} = \frac{116}{36}$.
   - Convert Back to Mixed Numbers: $\frac{116}{36} = 3\frac{8}{36}$.
   - Simplify: The greatest common factor of 36 and 8 is 4, so $8 \div 4 = 2$ and $36 \div 4 = 9$ and finally $3\frac{8}{36} = 3\frac{2}{9}$.

Multiplying and Dividing

Multiplying Fractions and Mixed Numbers:

1. **Convert Mixed numbers to Improper Fractions:** Convert each mixed number to an improper fraction.
2. **Multiply the Fractions:**
   - **Multiply the Numerators:** Multiply the numerators together to get the new numerator.
   - **Multiply the Denominators:** Multiply the denominators together to get the new denominator.
3. **Convert Back:** Convert the result back to a mixed number if needed.
4. **Simplify:** Simplify the resulting fraction if possible.

Dividing Fractions and Mixed Numbers:

1. **Convert Mixed Numbers to Improper Fractions:** Convert each mixed number to an improper fraction.
2. **Reciprocal:** Flip the second fraction (the divisor).
3. **Multiply:** Multiply the first fraction by the reciprocal of the second fraction.
4. **Convert Back:** Convert the result back to a mixed number if needed.
5. **Simplify:** Simplify the resulting fraction if possible.

Examples:

1) Multiply $\frac{5}{9}$ and $3\frac{1}{2}$:

   Solution:
   - Convert Mixed numbers to Improper Fractions: $3\frac{1}{2} = \frac{7}{2}$.
   - Multiply the Fraction: $\frac{5}{9} \times \frac{7}{2} = \frac{35}{18}$.
   - Convert Back: $\frac{35}{18} = 1\frac{17}{18}$.
   - Simplify: The fraction $1\frac{17}{18}$ is already in its simplest form.

2) Divide $1\frac{4}{5}$ by $5\frac{2}{3}$:

   Solution:
   - Convert Mixed numbers to Improper Fractions: $5\frac{2}{3} = \frac{17}{3}$ and $1\frac{4}{5} = \frac{9}{5}$.
   - Reciprocal: $\frac{17}{3} \rightarrow \frac{3}{17}$.
   - Multiply: $\frac{9}{5} \div \frac{17}{3} = \frac{9}{5} \times \frac{3}{17} = \frac{27}{85}$.
   - Convert Back: The result is a proper fraction.
   - Simplify: The fraction $\frac{27}{85}$ is already in its simplest form.

Mixed Operations

Mixed operations with fractions involve adding, subtracting, multiplying, and dividing fractions within the same problem. Here's a guide to handle these operations:

Steps to Perform Mixed Operations:

1. **Follow the Order of Operations:** Remember PEMDAS: Parentheses, Exponents, Multiplication and Division (from left to right), Addition and Subtraction (from left to right).
2. **Simplify Within Parentheses First:** Resolve any operations inside parentheses or brackets first.
3. **Convert Mixed Numbers to Improper Fractions:** For consistency, convert any mixed numbers to improper fractions before performing the operations.
4. **Find a Common Denominator for Addition/Subtraction:** When adding or subtracting fractions, convert them to have a common denominator.
5. **Multiply/Divide Fractions Directly:**
   - For multiplication, multiply the numerators and denominators.
   - For division, multiply by the reciprocal of the divisor.
6. **Simplify the Result:** Simplify the resulting fraction where possible.

Example:

Evaluate $2\frac{3}{4} + \frac{4}{5} \times \left(2 - \frac{1}{3}\right) \div \frac{8}{3}$:

Solution:

1. Parentheses: $2 - \frac{1}{3} = \frac{6}{3} - \frac{1}{3} = \frac{5}{3}$.
2. Convert Mixed Number: $2\frac{3}{4} = \frac{11}{4}$.
3. Multiplication and Simplifying: $\frac{4}{5} \times \frac{5}{3} = \frac{20}{15} = \frac{4}{3}$.
4. Division and Simplifying: $\frac{4}{3} \div \frac{8}{3} = \frac{4}{3} \times \frac{3}{8} = \frac{12}{24} = \frac{1}{2}$.
5. Addition: $\frac{11}{4} + \frac{1}{2} = \frac{11}{4} + \frac{2}{4} = \frac{13}{4}$.
6. Convert Back to Mixed Number: $\frac{13}{4} = 3\frac{1}{4}$.

Final answer: $2\frac{3}{4} + \frac{4}{5} \times \left(2 - \frac{1}{3}\right) \div \frac{8}{3} = 3\frac{1}{4}$.

Word Problems

Solving word problems involving fractions can be tackled step by step (Reading and Understanding, Translating, Performing the Operations, Simplifying and Checking Work) as previous chapters and you should note that, in some questions drawing a shape and doing the solution by that is helpful. In addition, it is very important to recognize the unit of each fraction to choose an appropriate approach. The following examples are apparently the same, but their solutions are different:

Examples:

1) First, Sara spent $\frac{1}{4}$ of her pocket money. Then, she bought a gift for her brother with $\frac{1}{3}$ of her whole initial money. What fraction of her money is left?
   Solution:
   1. Identify the Initial Fraction Spent: Sara spent $\frac{1}{4}$ of her pocket money initially.
   2. Determine the Total Fraction Spent: Add the fractions of money spent: $\frac{1}{4} + \frac{1}{3} = \frac{3}{12} + \frac{4}{12} = \frac{7}{12}$,
   3. Determine the Remaining Fraction: $1 - \frac{7}{12} = \frac{12}{12} - \frac{7}{12} = \frac{5}{12}$.

2) First, Sara spent $\frac{1}{4}$ of her pocket money. Then, she bought a gift for her brother with $\frac{1}{3}$ of the remaining money. What fraction of her money is left?
   Solution:
   The difference between this question and the previous one lies in the unit of second fraction. In question 1, the unit of $\frac{1}{3}$ is her whole money, but in this question, the unit of $\frac{1}{3}$ is her remaining money after first expenditure.
   1. Identify the Initial Fraction Spent: Sara spent $\frac{1}{4}$ of her pocket money initially.
   2. Determine the Remaining Money After the Initial Spend: She has $1 - \frac{1}{4} = \frac{4}{4} - \frac{1}{4} = \frac{3}{4}$,
   3. Calculate the Amount Spent on the Gift: She used $\frac{1}{3}$ of the remaining money for the gift: $\frac{1}{3} \times \frac{3}{4} = \frac{3}{12} = \frac{1}{4}$,
   4. Determine the Remaining Fraction of Her Money: Subtract the fraction spent on the gift from the remaining money: $\frac{3}{4} - \frac{1}{4} = \frac{2}{4} = \frac{1}{2}$.
   So, Sara has $\frac{1}{2}$ of her money left.
   ☑ For this question drawing a shape is an easier method for a solution:
   1. First show her first spent:
   2. Paint the $\frac{1}{3}$ of her remaining money:
   3. Now the final remaining money is white parts which is $\frac{2}{4}$ or $\frac{1}{2}$ of her whole money.

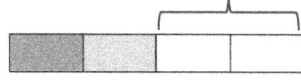

Worksheets

Converting between fraction and mixed numbers

🖎 Convert each mixed numbers to improper fractions and improper fractions to mixed numbers:

1) $3\frac{1}{4} =$
2) $4\frac{5}{6} =$
3) $1\frac{7}{10} =$
4) $5\frac{11}{20} =$
5) $9\frac{9}{9} =$

6) $\frac{17}{8} =$
7) $\frac{31}{5} =$
8) $\frac{63}{10} =$
9) $\frac{45}{4} =$
10) $\frac{39}{19} =$

Ordering fractions

🖎 Order following fractions from least to greatest:

1) $\frac{1}{5}, \frac{1}{3}, \frac{1}{11}$
2) $\frac{3}{10}, \frac{12}{5}, \frac{9}{15}$
3) $2\frac{5}{7}, 1\frac{1}{10}, \frac{32}{11}, \frac{14}{7}$
4) $\frac{24}{9}, \frac{29}{18}, \frac{17}{4}$
5) $\frac{5}{6}, \frac{3}{4}, \frac{8}{9}, \frac{2}{3}$

6) $\frac{13}{9}, \frac{13}{10}, \frac{13}{7}, \frac{13}{12}, \frac{13}{6}$
7) $1\frac{21}{30}, \frac{2}{3}, \frac{18}{5}, \frac{1}{2}, 2$
8) $\frac{12}{20}, \frac{9}{18}, \frac{17}{38}$
9) $\frac{25}{35}, \frac{1}{7}, \frac{36}{63}, \frac{16}{14}$
10) $6\frac{2}{9}, \frac{37}{5}, \frac{24}{8}, \frac{3}{20}, \frac{8}{9}$

Simplifying fractions

🖎 Simplify each fraction to its simplest form:

1) $\frac{5}{10} =$
2) $\frac{27}{36} =$
3) $\frac{32}{48} =$
4) $\frac{80}{160} =$
5) $\frac{63}{72} =$

6) $\frac{81}{90} =$
7) $\frac{28}{112} =$
8) $\frac{125}{200} =$
9) $\frac{270}{167} =$
10) $\frac{54}{126} =$

Least common denominator

🖎 Find the LCD of the following fractions:

1) $\frac{1}{10}, \frac{3}{5}$

6) $\frac{5}{16}, \frac{2}{24}$

2) $\frac{4}{15}, \frac{9}{20}$

3) $\frac{7}{24}, \frac{6}{12}$

4) $\frac{13}{40}, \frac{4}{30}$

5) $\frac{1}{35}, \frac{3}{70}$

7) $\frac{2}{18}, \frac{9}{15}$

8) $\frac{7}{36}, \frac{11}{24}, \frac{31}{8}$

9) $\frac{5}{30}, \frac{10}{18}, \frac{1}{9}$

10) $\frac{14}{180}, \frac{7}{27}, \frac{1}{100}, \frac{12}{18}$

Adding and subtracting

Do expressions:

1) $\frac{1}{2} + \frac{1}{7} =$

2) $2\frac{3}{8} + \frac{1}{4} =$

3) $3\frac{4}{5} - 1\frac{1}{8} =$

4) $\frac{5}{6} + 2\frac{2}{5} =$

5) $2\frac{9}{12} - \frac{3}{9} =$

6) $\frac{3}{5} + \frac{1}{10} + \frac{5}{6} =$

7) $1\frac{6}{10} - \frac{3}{20} + 2\frac{1}{15} =$

8) $2\frac{1}{48} + 1\frac{3}{16} + \frac{1}{6} =$

9) $9\frac{7}{10} - 5\frac{1}{8} - 1\frac{9}{20} =$

10) $3 - 1\frac{5}{7} + 2\frac{2}{7} =$

Multiplying and dividing

Do expressions:

1) $\frac{1}{5} \times \frac{15}{5} =$

2) $\frac{2}{3} \times \frac{7}{3} =$

3) $\frac{8}{9} \times \frac{9}{8} =$

4) $1\frac{2}{7} \times \frac{9}{4} =$

5) $\frac{3}{10} \times 2\frac{7}{5} =$

6) $\frac{12}{9} \div \frac{8}{7} =$

7) $\frac{4}{5} \div \frac{15}{5} =$

8) $6\frac{2}{3} \div \frac{5}{3} =$

9) $3\frac{1}{5} \div 4\frac{3}{5} =$

10) $1\frac{8}{20} \div 2\frac{3}{14} =$

Mixed operations

Evaluate operations:

1) $\frac{3}{9} \div \left(\frac{2}{5} \times \frac{1}{9}\right) =$

2) $2\frac{3}{4} \div \left(3\frac{4}{5} - 1\frac{1}{10}\right) =$

3) $\frac{2}{3} \div \frac{1}{4} + \frac{1}{3} \times \frac{1}{5} =$

4) $5 - 2\frac{1}{3} \times \frac{2}{3} + \frac{3}{4} =$

5) $5\frac{3}{10} + 3 \div \frac{1}{4} =$

6) $\frac{4}{5} \div 1\frac{2}{7} - \frac{2}{9} \times \frac{4}{16} =$

7) $\frac{6}{5} - \frac{9}{20} \div 2 \times \frac{1}{3} =$

8) $\frac{3}{4} \times \left(\left(\frac{2}{3} - \frac{1}{6}\right) \div \frac{5}{6}\right) =$

9) $\frac{10}{13} + 1\frac{5}{39} - \left(2 \times \frac{8}{13}\right) =$

10) $\frac{2}{5} \div \left(\frac{3}{5} \times 1\frac{1}{3} + \frac{3}{5}\right) =$

Word problem

Do word problems according to what you learned in this chapter:

1) Julia baked $\frac{3}{4}$ of a cake. She gave $\frac{1}{2}$ of what she baked to her friend and then ate $\frac{1}{3}$ of what was left. How much did Julia eat?

2) A recipe calls for $\frac{2}{3}$ cup of sugar, $\frac{1}{4}$ cup of butter, and $\frac{1}{2}$ cup of milk. If you want to triple the recipe, how much sugar, butter, and milk do you need in total?

3) John has $2\frac{1}{8}$ liters of paint. He uses $\frac{5}{10}$ of it for a project and then buys an additional $\frac{3}{5}$ liter of paint. How much paint does John have now?

4) Emily has $\frac{8}{10}$ of a yard of fabric. She needs to cut it into pieces that are $\frac{1}{4}$ of a yard each. How many pieces can she cut from the fabric?

5) A chemist has $3\frac{1}{3}$ liters of a solution. She needs $\frac{2}{5}$ liter for each experiment. How many experiments can she perform, and how much solution will be left over?

6) First, Sara read $\frac{4}{10}$ of her book. The next day, she read $\frac{1}{3}$ of the remaining book. If 120 pages of her book remained at the end of the second day, how many pages did her book have in total from the beginning?

7) A farmer planted wheat on $\frac{1}{2}$ of his land and then planted barley on $\frac{2}{3}$ of the remaining land. If his entire land was 60,000 square meters, how many square meters of wheat did he plant?

8) In a class, $\frac{9}{10}$ of the students are interested in sports. Of these, $\frac{2}{3}$ are interested in football and $\frac{2}{9}$ are interested in volleyball. The rest, who are 4 students, are interested in swimming.
   - How many students are there in total?
   - How many students are interested in football and volleyball?
   - How many students are not interested in any sports?

9) $\frac{3}{7}$ of the people in the park are adults. If there are 14 more children than adults, how many children are there in the park?

Answer of Worksheets

Converting between fraction and mixed numbers

1) $\frac{13}{4}$
2) $\frac{29}{6}$
3) $\frac{17}{10}$
4) $\frac{111}{20}$
5) 10

6) $2\frac{1}{8}$
7) $6\frac{1}{5}$
8) $6\frac{3}{10}$
9) $11\frac{1}{4}$
10) $2\frac{1}{19}$

Ordering fractions

1) $\frac{1}{11} < \frac{1}{5} < \frac{1}{3}$
2) $\frac{3}{10} < \frac{9}{15} < \frac{12}{5}$
3) $1\frac{1}{10} < \frac{14}{7} < 2\frac{5}{7} < \frac{32}{11}$
4) $\frac{29}{18} < \frac{24}{9} < \frac{17}{4}$
5) $\frac{2}{3} < \frac{3}{4} < \frac{5}{6} < \frac{8}{9}$

6) $\frac{13}{12} < \frac{13}{10} < \frac{13}{9} < \frac{13}{7} < \frac{13}{6}$
7) $\frac{1}{2} < \frac{2}{3} < 1\frac{21}{30} < 2 < \frac{18}{5}$
8) $\frac{17}{38} < \frac{9}{18} < \frac{12}{20}$
9) $\frac{1}{7} < \frac{36}{63} < \frac{25}{35} < \frac{16}{14}$
10) $\frac{3}{20} < \frac{8}{9} < \frac{24}{8} < 6\frac{2}{9} < \frac{37}{5}$

Simplifying fractions

1) $\frac{1}{2}$
2) $\frac{3}{4}$
3) $\frac{2}{3}$
4) $\frac{1}{2}$
5) $\frac{7}{8}$

6) $\frac{9}{10}$
7) $\frac{1}{4}$
8) $\frac{5}{8}$
9) $\frac{270}{167}$
10) $\frac{3}{7}$

Least common denominator

1) 10
2) 60
3) 24
4) 120
5) 70

6) 48
7) 90
8) 72
9) 90
10) 2,700

Adding and subtracting

1) $\frac{9}{14}$
2) $2\frac{5}{8}$
3) $2\frac{27}{40}$
4) $3\frac{7}{30}$
5) $2\frac{5}{12}$
6) $1\frac{8}{15}$
7) $3\frac{31}{60}$
8) $3\frac{3}{8}$
9) $3\frac{1}{8}$
10) $3\frac{4}{7}$

Multiplying and dividing

1) $\frac{3}{5}$
2) $1\frac{5}{9}$
3) 1
4) $2\frac{25}{28}$
5) $1\frac{1}{50}$
6) $\frac{7}{6}$
7) $\frac{4}{15}$
8) 4
9) $\frac{16}{23}$
10) $\frac{98}{155}$

Mixed operations

1) $7\frac{1}{2}$
2) $1\frac{1}{54}$
3) $2\frac{11}{15}$
4) $4\frac{7}{36}$
5) $17\frac{3}{10}$
6) $\frac{17}{30}$
7) $1\frac{1}{8}$
8) $\frac{9}{20}$
9) $\frac{2}{3}$
10) $\frac{2}{7}$

Word problem

1) $\frac{1}{8}$ of the cake.
2) Sugar: 2 cups, Butter: $\frac{3}{4}$ cups, Milk: $1\frac{1}{2}$ cups.
3) $1\frac{53}{80}$ liters.
4) $3\frac{1}{5}$ = 3 full pieces, $\frac{1}{5}$ of a piece left over.
5) 8 experiments and $\frac{1}{3}$ liter of solution left over.
6) 300 pages.
7) 30000 square meters.
8) 40 students in total, 24 students interested in football, 8 students interested in volleyball and 4 students not interested in any sports.
9) 56 children.

Chapter 5: Decimals

Topics that you will learn in this chapter:

- Ordering Decimals
- Converting Between Fractions and Decimal
- Converting Between Mixed Numbers and Decimals
- Adding and Subtracting Decimals
- Multiplying Decimals
- Dividing Decimals by Whole Numbers
- Dividing Decimals by Decimals
- Rounding Decimals
- Word Problems
- Worksheet

Ordering Decimals

Ordering decimals involves comparing their values to determine which is smaller or larger. Here's a step-by-step guide:

Compare Integer Parts: Start by comparing the integer parts of the numbers. The number with the smaller integer part is smaller overall.

Compare Decimal Parts: If the integer parts are the same, compare the decimal parts digit by digit.

Add Zeros (if necessary): To make comparison easier, add zeros to the end of shorter decimal numbers so they all have the same number of decimal places.

Examples:

1) Order decimal numbers from least to greatest:

   $3.45, 3.5, 0.78, 0.099, 2.9, 3.305, 4.1, 2.098, 1.56$

   Solution:

   1. Identify the Integer Parts: 3, 3, 0, 0, 2, 3, 4, 2, 1
   2. Order by Integer Parts: Smallest integer part to largest: 0, 1, 2, 3, 4
   3. Compare the Decimals with the Same Integer Part:
      - For decimals with integer part 0: $(0.099, 0.78) \, 0.099 < 0.78$.
      - For decimals with integer part 1: 1.56.
      - For decimals with integer part 2: $(2.098, 2.9) \, 2.098 < 2.9$.
      - For decimals with integer part 3: $(3.305, 3.45, 3.5) \, 3.305 < 3.45 < 3.5$.
      - For decimals with integer part 4: 4.1.
   4. Combine All Numbers in Order:
      $0.099 < 0.78 < 1.56 < 2.098 < 2.9 < 3.305 < 3.45 < 3.5 < 4.1$.

1) Order the decimals from largest to smallest:

   $2.35, 3.67, 0.456, 1.203, 2.5, 4.001, 0.678, 1.99, 3.07$

   Solution:

   1. Identify the Integer Parts: 2, 3, 0, 1, 2, 4, 0, 1, 3
   2. Order by Integer Parts: Largest integer part to smallest: 4, 3, 2, 1, 0
   3. Compare the Decimals with the Same Integer Part:
      - For decimals with integer part 4: 4.001
      - For decimals with integer part 3: $(3.67, 3.07) \, 3.67 > 3.07$
      - For decimals with integer part 2: $(2.35, 2.5) \, 2.5 > 2.35$
      - For decimals with integer part 1: $(1.203, 1.99) \, 1.99 > 1.203$
      - For decimals with integer part 0: $(0.456, 0.678) \, 0.678 > 0.456$
   4. Combine All Numbers in Order:
      $4.001 > 3.67 > 3.07 > 2.5 > 2.35 > 1.99 > 1.203 > 0.678 > 0.456$.

www.mathnotion.com

Fraction and Decimal

Converting Fractions to Decimals:

1. **If the denominator of fraction can be converted to a power of 10:**
   1. Write a fraction equal to that whose denominator is a power of ten.
   2. Write the numerator and specify the number of decimal places by the number of zeros in the denominator.
2. **If the denominator of fraction cannot be converted to a power of 10:** Divide the numerators by denominator (this specific problem will be thought in future sections and can be used for first kind of fractions too).

Converting Decimals to Fractions:

Count the number of digits after the decimal point and use that to determine the denominator. Then simplify the fraction if needed.

Examples:

1) Convert $\frac{2}{100}$ and $\frac{3}{4}$ to decimal:
   Solution:

   $\frac{2}{100}$:
   - Write a fraction equal to $\frac{2}{100}$ whose denominator is a power of ten: This fraction has already a denominator of 100.
   - Write the numerator (2) and the denominator (100) has 2 zeros, so the number of decimal places is 2: $\frac{2}{100} = 0.02$.

   $\frac{3}{4}$:
   - Write a fraction equal to $\frac{3}{4}$ whose denominator is a power of ten: 4 ×25=100 so $\frac{3}{4} = \frac{75}{100}$
   - Write the numerator (75) and the denominator (100) has 2 zeros, so the number of decimal places is 2: $\frac{75}{100} = 0.75$.

2) Convert 0.125 and 0.36 to fraction:
   Solution:

   0.125:
   - Write 0.125 as $\frac{125}{1000}$ (since there are three digits after the decimal, the denominator is 1000).
   - Simplify the $\frac{125}{1000} = \frac{1}{8}$ (by dividing the numerator and denominator by 125).

   0.36 :
   - Write 0.36 as $\frac{36}{100}$ (since there are two digits after the decimal, the denominator is 100).
   - Simplify the $\frac{36}{100} = \frac{9}{25}$ (by dividing the numerator and denominator by 4).

Mixed Numbers and Decimals

Converting between mixed numbers and decimals is quite like converting between fractions and decimals, here's a simple guide:

Converting Mixed Numbers to Decimals:

1. **Convert the Fraction Part:** The process is the same as the previous section!
2. **Add the Whole Number:** Add the decimal result of the fraction to the whole number part of the mixed number.

Converting Decimals to Mixed Numbers:

1. **Separate the Whole Number**: The whole number part of the decimal is the whole number part of the mixed number.
3. **Convert the Decimal Part to a Fraction:** The process is the same as the previous section!

Examples:

1) Convert $1\frac{3}{5}$ and $5\frac{11}{25}$ to decimal:

   Solution:

   $1\frac{3}{5}$:
   - Convert $\frac{3}{5}$ to a decimal: write a fraction equal to $\frac{3}{5}$ whose denominator is 10: $\frac{3}{5} = \frac{6}{10} = 0.6$.
   - Add the whole number: $1 + 0.6 = 1.6$.

   $5\frac{11}{25}$:
   - Convert $\frac{11}{25}$ to a decimal: write a fraction equal to $\frac{11}{25}$ whose denominator is 100: $\frac{11}{25} = \frac{44}{100} = 0.44$.
   - Add the whole number: $5 + 0.44 = 5.44$.

2) Convert 2.175 and 14.9 into mixed numbers:

   Solution:

   2.175 :
   - Separate the whole number: 2
   - Convert the decimal part to a fraction: $\frac{175}{1000}$.
   - Simplify the fraction: $\frac{175}{1000} = \frac{7}{40}$.
   - Combine: $2.175 = 2\frac{7}{40}$.

   14.8:
   - Separate the whole number: 14
   - Convert the decimal part to a fraction: $\frac{8}{10}$.
   - Simplify the fraction: $\frac{8}{10} = \frac{4}{5}$.
   - Combine: $14.8 = 14\frac{4}{5}$.

Adding and Subtracting Decimals

When adding or subtracting decimals, the process mirrors that of working with whole numbers, with the crucial step of ensuring the decimal points are properly aligned. Here's how you do it:

1. **Align the Decimal Points:** Write the numbers in a column, making sure the decimal points line up.
2. **Add Zeros if Necessary:** Add zeros to the end of any numbers to make sure they have the same number of decimal places.
3. **Add or Subtract Normally:** Start from the rightmost digit and move to the left, just like with whole numbers.
4. **Place the Decimal Point:** In the answer, place the decimal point directly below the other decimal points.

Example:

1) Add 3.56 and 1.789:

   Solution:

   Step 1: Write the numbers in a column:

      3.56
     +1.789
     _____

   Step 2: Add zeros to the end of 3.56:

      3.560
     +1.789
     _____

   Step 3: Add normally:

      3.560
     +1.789
     _____
      5349

   Step 4: Place the decimal point:

      3.560
     +1.789
     _____
      5.349

2) Subtract 5.981 from 8.19:

   Solution:

   Step 1: Write the numbers in a column:

      8.19
     - 5.981
     _____

   Step 2: Add zeros to the end of 3.56:

      8.190
     -5.981
     _____

   Step 3: Subtract normally:

      8.190
     -5.981
     _____
      2209

   Step 4: Place the decimal point:

      8.190
     -5.981
     _____
      2.209

Multiplying Decimals

Multiplying decimals works the same way as multiplying whole numbers with the difference that you need to be careful when placing the decimal point. Here are steps to multiply decimals:

1. **Ignore the Decimals Temporarily**: Treat the numbers as if they were whole numbers, ignoring the decimal points.
2. **Multiply Normally**: Perform the multiplication as you would with whole numbers.
3. **Count the Decimal Places**: Count the total number.

Examples:

1) Multiply 1.2 by 53.9:
   Solution:
   1. Ignore the Decimals: Multiply 12 and 539.
   2. Multiply:
      ```
        539
      × 12
      -----
       1078
      +5390
      -----
       6,468
      ```
   3. Count Decimal Place: 1.2 has 1 decimal place and 53.9 has 1 decimal place so, total decimal places are: 1 + 1 = 2.
   4. Place the Decimal Point: 64.68.

2) Multiply 0.08 by 0.235:
   Solution:
   1. Ignore the Decimals: Multiply 8 and 235.
   2. Multiply:
      ```
        235
      ×   8
      -----
       1,880
      ```
   3. Count Decimal Place: 0.08 has 2 decimal places and 0.235 has 3 decimal places so, total decimal places are: 2 + 3 = 5.
   4. Place the Decimal Point: You should note that the total digits of 1,880 is 4 but we have 5 decimal places so we must put a zero behind 1,880 and then place the decimal point: 0.01880.

3) Put decimal points for the product of multiplication:
   a) 0.8 × 0.02 = 16   b) 1.3 × 0.005 = 65   c) 0.0003 × 100 = 300
   Solution: According to "counting decimal place" rule:
   a) 0.016   b) 0.0065   c) 0.0300 = 0.03

www.mathnotion.com

Dividing Decimals by whole numbers

Dividing decimals by whole numbers with remainders follows a similar process to dividing whole numbers. Here's a step-by-step guide:

1. **Set up the Division**: Write the decimal number (the dividend) inside the division box and the whole number (the divisor) outside the box.
2. **Ignore the Decimal Point Initially**: Start by treating the decimal number as a whole number and perform division as usual. Divide each digit starting from the left.
3. **Place the Decimal Point in the Quotient**: Once you reach the decimal point in the dividend, place the decimal point directly above in the quotient.
4. **Continue Dividing:** Keep dividing as if you were working with a whole number.
5. **Check for Remainders**: If the division doesn't result in a whole number, continue by adding zeros after the decimal point and continue dividing until the remainder becomes zero or you reach the desired number of decimal places.

Example:

Divide 16.85 by 3 to three decimal places in quotient:

Solution:

Step 1: Divide 16 by 3

```
        5
   3) 16.85
      -15
      ___
        1
```

Step 2: Now we reach the decimal point in 16.85, place the decimal point above in the quotient (5).

```
       5.
   3) 16.85
      -15
      ___
        1
```

Step 3: Bring down the next digit (8) to get 18 and continue the division:

```
       5.61
   3) 16.85
      -15
      ___
       1.8
      -1.8
      ___
       0.5
      -0.3
      ___
       0.2
```

Step 4: You have 5.61 with a remainder of 0.2. Since you must have 3 decimal places in quotient, you can add a zero beside 0.2 making it 0.20 and continue the division:

```
       5.616
   3) 16.85
      -15
      ___
       1.8
      -1.8
      ___
       0.5
      -0.3
      ___
       0.20
      -0.18
      ____
       0.02
```

So, the result is approximately 5.616 with the remainder of 0.02.

Dividing Decimals by Decimals

Dividing decimals by decimals and finding a remainder involves similar steps to standard division, but you need to account for the decimal places. Here's how to do it step by step:

1. **Set up the Division**: Identify the two decimal numbers you want to divide.
2. **Eliminate Decimals**: Multiply both the dividend and divisor by the same power of 10 to convert them into whole numbers. It is important to know, If the dividend and divisor are multiplied by a number, the quotient does not change, but the remainder will be multiplied by that same number.
3. **Perform the Division:** Divide the resulting whole number.
   - ☑ If you want to do the division to several decimal places, put decimal point beside the quotient and continue the division by adding zero beside remainder.
4. **Find the Quotient and Remainder of Original Division**: The quotient is the same but for converting back to original remainder, you must divide the remainder by the number you'd been multiplied by dividend and divisor.

Example:

Divide 7.39 by 5.1 to one decimal place and determine the quotient and remainder:

Solution:

Step 1: Eliminate decimals: 7.39 has 2 decimal places and 5.1 have 1 decimal place so multiply by 100:

$7.39 \times 100 = 739$

$5.1 \times 100 = 510$

Step 2: Perform the division:

$$\begin{array}{r} 1 \\ 510{\overline{\smash{)}739}} \\ -510 \\ \hline 229 \end{array}$$

Step 3: Continue division by adding zero to reach a quotient with one decimal place:

$$\begin{array}{r} 1.4 \\ 510{\overline{\smash{)}739}} \\ -510 \\ \hline 229.0 \\ -204.0 \\ \hline 25.0 \end{array}$$

Step 4: Find the quotient and remainder of original division: As we multiplied 7.39 and 5.1 by 100, to get the original remainder we must divide the remainder by 100:

Original quotient: 1.4

Original remainder: $25 \div 100 = 0.25$

Multiply and Divide Decimals by the Power of Ten

Multiplying and dividing decimals by powers of ten (like 10, 100, 1000, etc.) is straightforward once you know the basic rules:

1. Multiplying Decimals by Powers of Ten:

When you multiply a decimal by a power of ten, move the decimal point to the right by as many places as there are zeros in the power of ten.

☑ If you move the decimal beyond the digits, fill in with zeros as needed.

2. Dividing Decimals by Powers of Ten:

When you divide a decimal by a power of ten, move the decimal point to the left by as many places as there are zeros in the power of ten.

☑ If there aren't enough digits to move left, add zeros in front of number.

Examples:
1) Multiply.
   - a) $0.28 \times 100 =$
   - b) $1.987 \times 10 =$
   - c) $36.1 \times 1000 =$
   - d) $0.006 \times 100 =$

   Solution:
   - a) Since 100 has two zeros, move the decimal two places to the right: $0.28 \times 100 = 28$
   - b) Since 10 has one zero, move the decimal one place to the right: $1.987 \times 10 = 19.87$
   - c) Since 1000 has three zeros, move the decimal three places to the right and as there are not enough digits after decimal point, fill it with 2 zeros. $36.1 \times 1000 = 36100$
   - d) Since 100 has two zeros, move the decimal two places to the right: $0.006 \times 100 = 0.6$

2) Divide.
   - a) $13.5 \div 10 =$
   - b) $621 \div 1000 =$
   - c) $8.072 \div 100 =$
   - d) $83.14 \div 100 =$

   Solution:
   - a) Since 10 has one zero, move the decimal one place to the left: $13.5 \div 10 = 1.35$
   - b) Since 1000 has three zeros, move the decimal three places to the left (we can suppose the decimal point to the right side of 621. $621 \div 1000 = 0.621$
   - c) Since 100 has two zeros, move the decimal two places to the left and as there are not enough digits before decimal point, fill it with 1 zero: $8.072 \div 100 = 0.08072$
   - d) Since 100 has two zeros, move the decimal two places to the left: $83.14 \div 100 = 0.8314$

Rounding Decimals

Rounding decimals is a process used to simplify a number while keeping it close to its original value. Here's how to round decimals step by step:

1. **Identify the Place Value**: Determine the place value you want to round to (e.g., tenths, hundredths, thousandths).
2. **Look at the Next Digit**: Check the digit immediately to the right of the place value you are rounding to:
   - If this digit is 0, 1, 2, 3, or 4, round down (leave the digit you're rounding as it is).
   - If this digit is 5, 6, 7, 8, or 9, round up (increase the digit you're rounding by 1).
3. **Adjust the Number**:
   - Keep all digits to the left of the rounding place value the same.
   - Change all digits to the right of the rounding place value to zero if rounding to a whole number or remove them if rounding to a decimal place.

Examples:

1) Round 3.456 to the nearest hundredth:

   Solution:
   - The hundredth place is 5 (3.456).
   - The next digit is 6 (greater than 5), so round up.
   - Result: 3.46

2) Round 7.842 to the nearest tenth:

   Solution:
   - The tenths place is 8 (7.842).
   - The next digit is 4 (less than 5), so round down.
   - Result: 7.8

3) Round 12.678 to the nearest whole number:

   Solution:
   - The whole number place is 12 (12.678).
   - The next digit is 6 (greater than 5), so round up.
   - Result: 13

4) Determine the place value that each number has been rounded by:

   a) $14.899 \approx 14.9$ b) $537.26 \approx 537$ c) $0.1877 \approx 0.188$

   Solution:

   According to the rounding rules:

   a) Tenths or hundredths   b) Whole number   c) Thousandths

Word Problems

Solving word problems involving decimals and operations on them requires a systematic approach. The general approach is the same as it was mentioned in the first chapter (Reading and Understanding, Translating, Performing the Operations, Simplifying and Checking Work). In addition, identify the decimals and operations and sometimes estimating the answer is helpful:

Example:

David is driving on a road trip, and his car's fuel efficiency is 23.6 miles per gallon. His fuel tank holds 15.5 gallons of gas. If he fills up his tank and drives until the tank is empty, how much will David spend on gas if the price is $4.25 per gallon, and how many miles can he drive on a full tank?

Solution:

Step 1: **Reading and Understanding the Problem**:

- David's car's fuel efficiency is 23.6 miles per gallon.
- The fuel tank holds 15.5 gallons of gas.
- The price of gas is $4.25 per gallon.

Step 2: **Identify the Operation:**

To calculate the total distance David can drive on a full tank, multiply the fuel efficiency (miles per gallon) by the number of gallons in the tank, and to find out how much David will spend, multiply the number of gallons in the tank by the price per gallon.

Step 3: **Perform the Operations:**

23.6 miles/gallon × 15.5 gallons = 365.8 miles.

15.5 gallons × 4.25 $/gallon = 65.875 dollars.

Final answer:

David can drive 365.8 miles on a full tank.

He will spend $65.88 to fill up his tank.

www.mathnotion.com

Worksheets

Ordering Decimals

🖎 Order decimals from least to greatest:

1) 0.21, 0.139, 0.3, 0.301, 0.198   —, —, —, —, —

2) 0.09, 1.1, 0.145, 0.2, 1.0008   —, —, —, —, —

3) 31.2, 4.02, 2.984, 32.01, 40.2   —, —, —, —, —

4) 18.75, 14.999, 21.1, 19.013, 22   —, —, —, —, —

5) 1.39, 0.7999, 2.1, 0.8, 2.054   —, —, —, —, —

🖎 Write the correct comparison symbol (<, = or >):

6) 7.001 ☐ 7.01   9) 1.39 ☐ 1.3921

7) 81.2 ☐ 81.189   10) 264.2 ☐ 26.42

8) 0.0236 ☐ 0.1

Fraction and Decimal

🖎 Convert fractions to decimals:

1) $\frac{1}{2}$ 2) $\frac{2}{5}$ 3) $\frac{17}{20}$ 4) $\frac{3}{25}$ 5) $\frac{5}{8}$

🖎 Convert decimals to fractions:

6) 0.32   7) 1.4   8) 0.981   9) 36.5   10) 12.78

Mixed Numbers and Decimals

🖎 Convert mixed numbers to decimals:

1) $3\frac{9}{50}$ 2) $1\frac{1}{4}$ 3) $10\frac{18}{1000}$ 4) $6\frac{3}{100}$ 5) $14\frac{4}{16}$

🖎 Convert decimals into mixed numbers:

6) 1.25   7) 5.912   8) 10.8   9) 123.4   10) 9.356

Adding and Subtracting Decimals

🖎 Do expressions:

1) 0.23 + 1.1 =   6) 2.104 − 1.9 =

2) 2.05 + 0.098 =   7) 1.895 − 0.6 =

www.mathnotion.com

3) $3.9 + 1.361 =$

4) $18.3 + 20.023 =$

5) $17 + 3.21 =$

8) $2.97 - 1.361 =$

9) $9 - 5.314 =$

10) $3 - 2.196 =$

Multiplying Decimals

✎ Multiply.

1) $0.8 \times 1.3 =$

2) $14.2 \times 0.9 =$

3) $38.02 \times 0.01 =$

4) $12 \times 0.4 =$

5) $13.7 \times 14 =$

6) $0.003 \times 0.02 =$

7) $19.02 \times 60 =$

8) $78.6 \times 500 =$

9) $8.41 \times 0.06 =$

10) $7.16 \times 5.3 =$

Dividing Decimals by whole numbers

✎ Divide.

1) $3.4 \div 2 =$

2) $9.63 \div 3 =$

3) $0.64 \div 8 =$

4) $2.5 \div 5 =$

5) $0.048 \div 12 =$

6) $1.44 \div 7 =$

7) $8.751 \div 13 =$

8) $23.8 \div 10 =$

9) $652.1 \div 9 =$

10) $5.23 \div 9 =$

Dividing Decimals by Decimals

✎ Do divisions.

1) $1.32 \div 0.06 =$

2) $31.5 \div 6.3 =$

3) $0.123 \div 0.041 =$

4) $98.3 \div 56.4 =$

5) $0.048 \div 0.012 =$

6) $8.51 \div 6.1 =$

7) $19.24 \div 3.7 =$

8) $7.61 \div 8.3 =$

9) $98.5 \div 9.4 =$

10) $0.31 \div 0.2 =$

Multiply and Divide Decimals by the Power of Ten

✎ Multiply expressions.

1) $0.27 \times 100 =$

2) $3.784 \times 100 =$

6) $18.6 \div 10 =$

7) $81.97 \div 100 =$

3) 0.11 × 10 =

4) 12.365 × 1000 =

5) 0.014 × 10 =

8) 7.91 ÷ 1000 =

9) 0.16 ÷ 10 =

10) 7.126 ÷ 1000 =

Rounding Decimals

✍ Round each number to the underlined place value:

1) 18.<u>3</u>69 =

2) 0.0<u>0</u>77 =

3) 98<u>1</u>.6 =

4) <u>2</u>1.57 =

5) 1<u>8</u>7.14 =

6) 1<u>1</u>.578 =

7) 9.8<u>1</u>4 =

8) 14.3<u>7</u>8 =

9) <u>9</u>.314 =

10) 5.02<u>58</u>7 =

Word Problems

✍ Do word problems according to what you learned in this chapter:

1) A recipe for pancakes requires 1.75 cups of flour, 0.5 cups of sugar, and 0.25 cups of butter. If Omar wants to double the recipe, how much of each ingredient will he need in total?

2) Maria buys 2.5 pounds of apples at $1.80 per pound and 1.75 pounds of oranges at $2.30 per pound. How much does she spend in total?

3) If the thickness of a book with 250 pages is 3.16 cm, how many centimeters is the thickness of one sheet of the book?

4) Leo's family is going on a road trip. They drive 145.6 miles on 8 gallons of gas. How many miles per gallon did the car use?

5) Emma bought 3.75 meters of ribbon. She used 1.25 meters for a project and decided to cut the remaining ribbon into 5 equal pieces. How long is each piece?

6) A car travels 150.6 miles on 12.5 gallons of gasoline. What is the car's fuel efficiency in miles per gallon?

7) Maria bought 3.5 meters of cloth. She used $\frac{2}{5}$ of it to make a dress and 0.75 meters to make a scarf. How much cloth does she have left?

8) A car travels 2.75 miles for every $\frac{1}{3}$ gallon of fuel. If the car's fuel tank can hold 15.5 gallons, how far can the car travel on a full tank?

Answer of Worksheets

Ordering Decimals

1) 0.139 < 0.198 < 0.21 < 0.3 < 0.301.
2) 0.09 < 0.145 < 0.2 < 1.0008 < 1.1
3) 2.984 < 4.02 < 31.2 < 32.01 < 40.2
4) 14.999 < 18.75 < 19.013 < 21.1 < 22
5) 0.7999 < 0.8 < 1.39 < 2.054 < 2.1
6) 7.001 $\boxed{<}$ 7.01
7) 81.2 $\boxed{>}$ 81.189
8) 0.0236 $\boxed{<}$ 0.1
9) 1.39 $\boxed{<}$ 1.3921
10) 264.2 $\boxed{>}$ 26.42

Converting Between Fraction and Decimal

1) 0.5
2) 0.4
3) 0.85
4) 0.12
5) 0.625
6) $\frac{8}{25}$
7) $\frac{7}{5}$
8) $\frac{981}{1000}$
9) $\frac{73}{2}$
10) $\frac{639}{50}$

Converting Between Mixed Numbers and Decimals

1) 3.18
2) 1.25
3) 10.018
4) 6.03
5) 14.25
6) $\frac{5}{4}$
7) $\frac{2956}{500}$
8) $\frac{54}{5}$
9) $\frac{617}{5}$
10) $\frac{2339}{250}$

Adding and Subtracting Decimals

1) 1.33
2) 2.148
3) 5.261
4) 38.323
5) 20.21
6) 0.204
7) 1.295
8) 1.609
9) 3.686
10) 0.804

Multiplying Decimals

1) 1.04
2) 12.78
3) 0.3802
4) 4.8
5) 191.8
6) 0.00006
7) 1,141.2
8) 39,300
9) 0.5046
10) 37.948

Dividing Decimals by whole numbers

1) 1.7
6) ≈ 0.205

www.mathnotion.com

2) 3.21
3) 0.08
4) 0.5
5) 0.004

7) ≈ 0.673
8) 2.38
9) ≈ 72.45
10) ≈ 0.581

Dividing Decimals by Decimals

1) 22
2) 5
3) 3
4) 1.742 and R:0.0512
5) 4

6) 1.395 and R:0.0005
7) 5.2
8) 0.916 and R:0.0072
9) 10.478 and 0.0068
10) 1.55

Multiply and Divide Decimals by the Power of Ten

1) 27
2) 378.4
3) 1.1
4) 12365
5) 0.14

6) 1.86
7) 0.8197
8) 0.00791
9) 0.016
10) 0.007126

Rounding Decimals

1) 18.4
2) 0.01
3) 982
4) 20
5) 190

6) 12
7) 9.81
8) 14.38
9) 9
10) 5.0259

Word Problems

1) Flour: 3.5 cups
   Sugar: 1 cup
   Butter: 0.5 cups
2) 8.525 dollars.
3) 0.02528 cm.
4) 18.2 miles per gallon.

5) 0.5 meters.
6) 12.048 miles per gallon.
7) 1.35 meters.
8) 127.875 miles.

Chapter 6: Ratios, Proportions and Percents

Topics that you will learn in this chapter:

- Writing and Identifying Ratios
- Simplifying Ratios
- Equivalent Ratios and Rates
- Proportional in Tables and Graphs
- Ratio and Rates Word Problems
- Understanding Percent
- Converting Between Fractions, Mixed Numbers and Percent
- Converting Between Percent and Decimals
- Finding Percents of Numbers
- Percent Word Problems
- Discount, Tax and Tip
- Worksheets
- Answer of Worksheets

Writing and Identifying Ratios

Ratios are a way to compare two or more quantities. They show how much of one thing there is compared to another. Ratios can be written in three ways:

1. Using the colon symbol (:): 3:2
2. As a fraction: $\frac{3}{2}$
3. Using the word "to": 3 to 2

In general, when we want to write ratios, we write the first mentioned quantities in numerator and the second one in denominator.

Examples:

1) According to the shape below write wanted ratios:

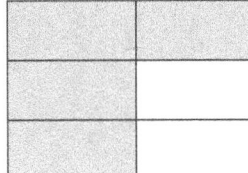

a) The ratio of gray parts to whole parts.
b) The ratio of white parts to whole parts.
c) The ratio of gray parts to white parts.
d) The ratio of white parts to gray parts.

Solution:

We must write the first quantities for the numerator and the second one for denominator:

a) $\frac{4}{6}$ b) $\frac{2}{6}$ c) $\frac{4}{2}$ d) $\frac{2}{4}$

2) If the ratio of apples to the oranges is $\frac{3}{7}$ and the ratio of oranges to lemons is $\frac{7}{10}$, what is the ratio of apples to lemons?

Solution:

$\frac{apples}{oranges} = \frac{3}{7}$ and $\frac{oranges}{lemons} = \frac{7}{10}$, the quantity of oranges in both ratios is 7 so the ratio of apples to lemons is $\frac{3}{10}$.

3) If the ratio of Jason's money to Maya's money is $\frac{5}{3}$ and the ratio of Jason's money to Emma's money is $\frac{15}{8}$, what is the ratio of Emma's money to Maya's money?

Solution:

$\frac{Jason's\ money}{Maya's\ money} = \frac{5}{3}$ and $\frac{Jason's\ money}{Emma's\ money} = \frac{15}{8}$. Jason's money is common in both ratios, first we must equalize the Jason's money in both ratios: $\frac{Jason's\ money}{Maya's\ money} = \frac{5}{3} = \frac{15}{9}$, so $\frac{Emma's\ money}{Maya's\ money} = \frac{8}{9}$.

Simplifying Ratios

Simplifying ratios are similar to simplifying fractions. Here's a step-by-step guide:

Steps to Simplify Ratios:

1. **Write the Ratio**: Identify the two quantities in the ratio and write them as a fraction.
2. **Find the Greatest Common Divisor (GCD):** Determine the largest number that divides both quantities without leaving a remainder. This process is just like finding GCF in chapter 3 (Find prime factorization of each number then identify common prime factors and finally multiply common prime factors
3. **Divide Both Quantities by the GCD:** Divide both the numerator and the denominator of the fraction by the GCD.
4. **Write the Simplified Ratio:** The resulting fraction gives you the simplified ratio.

Examples:

1) Simplify the ratio 18:24.

   Solution:
   1. Write the Ratio as a Fraction: $\frac{18}{24}$.
   2. Find the GCD of 18 and 24: $18 = 2 \times 3 \times 3$ and $24 = 2 \times 2 \times 2 \times 3$, so, the GCD = $2 \times 3 = 6$.
   3. Divide Both Quantities by the GCD: $18 \div 6 = 3$ and $24 \div 6 = 4$.
   4. Write the Simplified Ratio: The simplified ratio is $\frac{18}{24} = \frac{3}{4}$.

2) Simplify the ratio 66: 123.

   Solution:
   1. Write the Ratio as a Fraction: $\frac{66}{123}$.
   2. Find the GCD of 66 and 123: $66 = 2 \times 3 \times 11$ and $123 = 3 \times 41$, so the GCD = 3
   3. Divide Both Quantities by the GCD: $66 \div 3 = 22$ and $123 \div 3 = 41$.
   4. Write the Simplified Ratio: The simplified ratio is $\frac{66}{123} = \frac{22}{41}$.

3) Which ratios are equal with the ratio of 5:9?
   a) 30:45   b) 15:18   c) 20:36   d) 35:54

   Solution:

   Simplify all ratios using the process above:

   a) $\frac{30}{45} = \frac{2}{3}$ b) $\frac{15}{18} = \frac{5}{6}$ c) $\frac{20}{36} = \frac{5}{9}$ d) $\frac{35}{56} = \frac{5}{8}$

   So, the answer is c.

www.mathnotion.com

Equivalent Ratios and Rates

Finding equivalent ratios and rates involves creating two sets that express the same relationship. For equivalizing ratios and rates do these steps:
1. **Identify the Original Ratio:** Write down the given ratio.
2. **Multiply or Divide Both Terms by the Same Number:** To find an equivalent ratio, multiply or divide both parts of the ratio by the same non-zero number.

☑ It is very important to know the difference between ratios and rates. Here are two key differences:
- **Units**: Ratios compare similar units, while rates compare different units.
- **Context**: Ratios are often used in contexts where both quantities are part of the same whole. Rates measure how one quantity changes relative to another.

Examples:

1) Find equivalent ratios for 2:3.
   Solution:
   - Multiply both parts by 2: $2 \times 2 : 3 \times 2 = 4:6$ or $\frac{2 \times 2}{3 \times 2} = \frac{4}{6}$ so, $\frac{2}{3} = \frac{4}{6}$
   - Multiply both parts by 3: $2 \times 3 : 3 \times 3 = 6:9$ or $\frac{2 \times 3}{3 \times 3} = \frac{6}{9}$ so, $\frac{2}{3} = \frac{6}{9}$
   - Equivalent ratios: 2:3, 4:6, 6:9.

2) Find equivalent rates for $\frac{5 \, miles}{1 \, hour}$.
   Solution:
   - Multiply both by 2: $\frac{5 \times 2 \, miles}{1 \times 2 \, hour} = \frac{10 \, miles}{2 \, hour}$
   - Multiply both by 3: $\frac{5 \times 3 \, miles}{1 \times 3 \, hour} = \frac{15 \, miles}{3 \, hour}$
   - Equivalent rates: $\frac{5 \, miles}{1 \, hour}, \frac{10 \, miles}{2 \, hour}, \frac{15 \, miles}{3 \, hour}$

3) Fill in the blanks by calculating each equivalence:
   a) 3:8=☐:48   b) 6:7=12:☐   c) ☐:9= 5:3   d) 4:☐=16:28

   Solution: For each equivalent we should do the steps:
   1. **Set up the equivalence**
   2. **Identify the relationship between the known quantities**
   3. **Apply the same multiplication or division to the blank**

   a) $3:8 = \frac{3 \times 6}{8 \times 6} = \frac{18}{48} = 18:48$
   b) $6:7 = \frac{6 \times 2}{7 \times 2} = \frac{12}{14} = 12:14$
   c) $15:9 = \frac{15}{9} = \frac{5 \times 3}{3 \times 3} = 5:3$
   d) $4:7 = \frac{4}{7} = \frac{16 \div 4}{28 \div 4} = 16:28$

Proportion in Tables and Graphs

Proportion means that the two ratios are equal. It's a way of stating that two sets of quantities have the same relative size or ratio.

Showing Proportions in Tables:

In tables, if the relationship between two quantities is proportional, the ratio remains consistent across all data points.

Finding Proportions Using Tables:

1. **Set Up the Table:**
   - List the quantities being compared.
   - Check if the ratio between the quantities is constant.
2. **Calculate Ratios:** Divide one quantity by the other for each pair to see if the ratio is consistent.
3. **Determine Proportionality:** If all the ratios are equal, the quantities are proportional.

Showing Proportions in Graphs:

On graphs, a proportional relationship is represented by a straight line passing through the origin (0,0). This means as one quantity increases, the other increases at a constant rate.

Finding Proportions Using Graphs:

1. **Plot the Data Points:** Place each pair of quantities on a graph with x on the horizontal axis and y on the vertical axis.
2. **Draw the Line**: Connect the points or observe if they align in a straight line.
3. **Check for Proportionality**: If the points form a straight line passing through the origin (0,0), the quantities are proportional.

Example

If there is a proportional relationship between x and y complete the table and graph the data from table:

x	2	4	6	12
y	3			

Solution:

Since the relationship between x and y is proportional so we can write equivalent ratio for x and y:

$2 \times 2 : 3 \times 2 \rightarrow 4 : 6$,

$2 \times 3 : 3 \times 3 \rightarrow 6 : 9$,

$2 \times 4 : 3 \times 4 \rightarrow 8 : 12$

x	2	4	6	8
y	3	6	9	12

graph plot points:
(2,3), (4,6), (6,9), (8, 12) and draw the line:

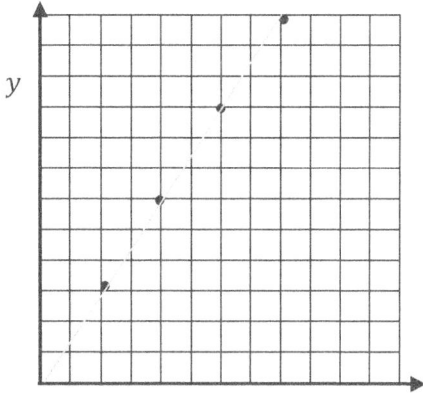

Ratio and Rates Word Problems

Solving ratio and rates word problems involves understanding the relationship between the quantities and using mathematical operations to find the solution. Here's a guide:

Steps to Solve Ratio/Rates Word Problems:

1. **Understand the Problem**: Read the problem carefully to identify the quantities being compared.
2. **Set Up the Ratio/Rate**: Write down the given ratio or rate (e.g., miles per hour, cost per item) in the problem.
3. **Use the Ratio/Rate:**
   - For ratio: Apply the ratio to the quantities given to find the missing value.
   - For rate: Multiply or divide using the rate to find the unknown quantity.
4. **Simplify if Necessary:** Simplify the ratio to make calculations easier.
5. **Convert Units if Necessary (For Rates):** Ensure all units are consistent.

Examples:

1) In a classroom, the ratio of boys to girls is 4:5. If there are 20 boys, how many girls are there?
   Solution:
   1. Understand the Problem:
      - Ratio of boys to girls: 4:5
      - Number of boys = 20
   2. Set Up the Ratio: $\frac{boys}{girls} = \frac{4}{5}$
   3. Use the Ratio: $\frac{20}{x} = \frac{4}{5}$
   4. Find Missing Value: $\frac{20}{x} = \frac{4 \times 5}{5 \times 5} = \frac{20}{25}$.

   So, the answer is 25.

2) A factory produces 120 widgets in 8 hours. At this rate, how many widgets can the factory produce in 12 hours?
   Solution:
   1. Understand the Problem:
      - Rate of production: $\frac{120\ widgets}{8\ hours}$
      - Time = 12 hours
   2. Set Up the Rate: $\frac{120\ widgets}{8\ hours}$ =15 widgets per hour
   3. Use the Rate: Calculate the total widgets in 12 hours: 15 widgets per hour ×12 hour= 180 widgets.

   So, the factory can produce 180 widgets in 12 hours.

www.mathnotion.com

Converting Between Fractions, Mixed Numbers and Percent

Percent means "per hundred" and is a way to express a number as a fraction of 100. It's used to compare proportions and understand parts of a whole easily.

Converting Fractions to Percents: Divide the numerator (top number) by the denominator (bottom number) to convert it to decimal and then multiply the decimal by 100.

☑ If it's possible, write an equivalent fraction with a denominator of 100, so the numerator is the exact percent.

Convert Percents to Fractions: Write the percent as a fraction over 100 and then simplify the fraction to its simplest form (Find GCD and divide both numerator and the denominator).

Convert Mixed Numbers to Percents: Convert the mixed number to an improper fraction, then divide the numerator by the denominator and finally multiply the decimal by 100.

☑ If it's possible, after converting decimal to improper fraction, write an equivalent fraction with a denominator of 100, so the numerator is the exact percent.

Convert Percents to Mixed Numbers: Write the percent as a fraction over 100, then simplify the fraction to its simplest form and convert it to a mixed number.

Examples:

1) Convert $\frac{11}{20}$ to percentage.

   Solution:
   It is simpler to write an equivalent fraction with a denominator of 100 because $20 \times 5 = 100$ so: $\frac{11}{20} = \frac{55}{100}$, and the numerator (55) is the exact percent: $\frac{55}{100} = 55\%$.

2) Convert 32% to fraction:

   Solution:
   Write the 32% as a fraction: $32\% = \frac{32}{100}$ and simplify it to its simplest form (GCD=4): $\frac{32}{100} = \frac{8}{25}$.

3) Convert $2\frac{5}{16}$ to percentage:

   Solution:
   Convert $2\frac{5}{8}$ to improper number: $2\frac{5}{8} = \frac{21}{8}$, now divide 21 by 8: $21 \div 8 = 2.625$ and then multiply 2.625 by 100: $2.625 \times 100 = 262.5\%$

4) Convert 175 % to mixed numbers:

   Solution:
   Write the 175% as a fraction: $175\% = \frac{175}{100}$ and simplify it to its simplest form (GCD=25): $\frac{175}{100} = \frac{7}{4} = 1\frac{3}{4}$.

```
     2.625
   8)21
    -16
     5.0
    -4.8
     0.20
    -0.16
     0.40
    -0.40
      0
```

www.mathnotion.com

Converting Between Percent and Decimals

When you see a number with a % symbol, it means "out of 100." Understanding this will help you easily convert between percents and decimals.

Converting Percent to Decimal: Remove the percent sign and then divide by 100 or move the decimal point two places to the left.

Converting Decimal to Percent: Multiply the decimal by 100 (or move the decimal point two places to the right) and then add the percent sign.

Just remember: **Percent ↔ Decimal** is all about moving the decimal point 2 places or dividing/multiplying by 100.

Examples:

1) Convert percents to decimals:

   a) 93 % b) 24% c) 184% d) 12.5%

   Solution:

   a) Remove the percent sign (93) and move the decimal point two place to the left (93 doesn't have decimal point so we suppose the decimal point right side of 3 (93.0); 93% = 0.93
   b) Just like part we have: 24% = 0.24
   c) Just like part we have: 184% = 1.84
   d) This number has the decimal point, so to convert it to decimal move the decimal point two places to the left: 12.5% = 0.125

2) Convert the decimals to percents:

   a) 0.25 b) 0.18 c) 1.48 d) 23.4

   Solution:

   a) Multiply 0.25 by 100 or move the decimal point two place to the right: $0.25 \times 100 = 25$ and put the percent sign: 25%
   b) Just like part a: $0.18 \times 100 = 18$ putting the percent sign: 18%
   c) Just like part a: $1.48 \times 100 = 148$ putting the percent sign: 148%
   d) Just like part a: $23.4 \times 100 = 2,340$ putting the percent sign: 2340%

www.mathnotion.com

Finding Percents of Numbers

Steps to Find the Percent of a Number:

- **Convert the Percent to a Decimal:** Divide the percent by 100.
- **Multiply the Decimal by the Number:** Multiply the converted decimal by the number you want to find the percent of.

Examples:

1) Find the answer to each part:
   a) 30% of 150
   b) 15% of 80
   c) 120% of 1.5
   d) 40% of $\frac{6}{5}$

   Solution:
   a) Convert 30% to decimal and then multiply by 150:
   $$30 \div 100 = 0.30 \to 0.30 \times 150 = 45$$

   b) Convert 15% to decimal and then multiply by 80:
   $$15 \div 100 = 0.15 \to 0.15 \times 80 = 12$$

   c) Convert 120 % to decimal and then multiply by 1.5:
   $$120 \div 100 = 1.2 \to 1.2 \times 1.5 = 1.8$$

   d) Convert 40% to decimal and then multiply by $\frac{6}{5}$ (we should convert $\frac{6}{5}$ to decimal):
   $$\frac{6}{5} = 6 \div 5 = 1.2 \to 40 \div 100 = 0.40 \to 0.40 \times 1.2 = 0.48$$

2) If 40% of a number is 800, how much is the number?

   Solution:

   This problem is the opposite of the previous problem, First convert the 40% to decimal:

   40% = 40 ÷ 100 = 0.40 and we have: 0.40 × ☐ = 800,

   To get the number we must divide 800 by 0.40:

   ☐ = 800 ÷ 0.40 = 2,000, so the number is 2,000.

3) What percentage of 36 is equal to 9?

   Solution:

   By using similar methods in previous problems, we have:

   ☐ × 36 = 9, to get the percent we must divide 9 by 36:

   ☐ = 9 ÷ 36 = 0.25, now convert 0.25 to a percent by multiplying 0.25 by 100:

   0.25 × 100 = 25%.

www.mathnotion.com

Discount, Tax and Tip

Working on word problems about discounts, taxes, and tips involves some straightforward math, but understanding the context is key. Here's how to break it down step-by-step:

Discount Problems:

- **Find the Discount Amount:** Convert the discount percent to decimal and then multiply the original price by decimal.
- **Subtract the Discount from the Original Price.**

Tax Problems:

- **Find the Tax Amount:** Convert the tax percent to decimal and then multiply the price by decimal.
- **Add the Tax to the Original Price.**

Tip Problems:

- **Find the Tip Amount:** Convert the tip percent to decimal and then multiply the meal cost by decimal.
- **Add the Tip to the Original Amount.**

Examples:

1) A dress originally cost $60. It's on sale for 20% off. How much do you pay?
   Solution:
   - Find the Discount Amount: $20 \div 100 = 0.20$, $60 \times 0.20 = 12$
   - Subtract the Discount from the Original Price: $60 - 12 = 48$

   You pay $48.

2) You buy a pair of shoes for $50. The sales tax is 8%. How much is the total cost?
   Solution:
   - Find the Discount Amount: $8 \div 100 = 0.08$, $50 \times 0.08 = 4$
   - Add the Tax to the Original Price: $50 + 4 = 54$

   The total cost is $54.

3) You have a meal that costs $30. You want to give a 15% tip. How much is the tip, and what is the total amount you pay?
   Solution:
   - Find the Tip Amount: $15 \div 100 = 0.15$, $30 \times 0.15 = 4.50$
   - Add the Tip to the Original Amount: $30 + 4.50 = 34.50$

   The total you pay is $34.50.

www.mathnotion.com

Word Problems

Similar to previous chapters and here for solving word problems involving percents we should follow these steps:

1. **Understand the Problem:** Read the problem carefully to identify the quantities involved and what is being asked.
2. **Convert the Percent to a Decimal:** Divide the percent by 100.
3. **Set Up the Equation:** Use the decimal to set up an equation involving the given quantity.
4. **Solve the Equation:** Multiply the decimal by the quantity to find the result.
5. **Answer the Question:** Ensure your answer makes sense in the context of the problem.

Example:

A laptop is originally priced at $1,200. During a holiday sale, the store offers a 15% discount. After the discount, an additional 8% sales tax is applied. What is the final price of the laptop?

Solution:

1. Find the Discount Amount:
    - Convert 15% to a decimal: 15% = 15 ÷ 100 = 0.15
    - Multiply the original price by the discount rate: 1,200 × 0.15 = 180
    - The discount amount is $180.
2. Subtract the Discount from the Original Price:
    - Subtract the discount from the original price: 1,200 - 180 = 1,020
    - The price after the discount is $1,020.
3. Calculate the Sales Tax:
    - Convert 8% to a decimal: 8 ÷ 100 = 0.08
    - Multiply the discounted price by the sales tax rate: 1,020 × 0.08 = 81.6
    - The sales tax amount is $81.60.
4. Add the Sales Tax to the Discounted Price:
    - Add the sales tax to the discounted price: 1,020 + 81.6 = 1,101.6

The final price of a laptop is $1,101.60.

Worksheets

Writing and Identifying Ratios

✍ According to the width and length of the rectangle below find the ratios:

1) The ratio of width to length
2) The ratio of length to its perimeter
3) The ratio of it's perimeter to width
4) The ratio of its area to its perimeter
5) The ratio of half of its perimeter to sum of its length and width

6 cm
10 cm

✍ If the ratio of oranges to apples is $\frac{6}{7}$, the ratio of apples to lemons is $\frac{14}{5}$ and the ratio of lemons to peaches is $\frac{10}{3}$, find following ratios:

6) Ratio of apples to peaches
7) Ratio of oranges to lemons
8) Ratio of peaches to oranges
9) Ratio of lemons and apples to oranges
10) Ratio of peaches and oranges to apples

Simplifying Ratios

✍ Simplify ratios to their simplest form:

1) $45:36$
2) $12:48$
3) $81:27$
4) $32:56$
5) $14:49$
6) $84:78$
7) $125:145$
8) $90:270$
9) $141:708$
10) $123:504$

Equivalent Ratios and Rates

✍ Find three different equivalent ratios for the ratios:

1) 2:3   2) 6:9   3) 8:7   4) 14:20   5) 24:15

✍ Fill in the blanks by calculating each equivalence:

6) 2:7=☐:14   7) 8:6= 32:☐   8) ☐:36 = 5:6   9) 21:24=☐:16   10) 12:☐ = 8:12

Proportion in Tables and Graphs

✍ Check the tables below to see if they are proportional or not:

1)

X	4	8	12
y	5	10	15

2)

X	8	24	16
y	9	18	27

3)

X	15	12	9
y	5	4	3

4)

X	36	15	6
y	18	7.5	3

☙ Complete proportional tables.

5)

X	3	9	
y	10		40

6)

X		3	30
y	16	4	

7)

X	35	7	
y	25		10

8)

X		1	4
y	24		6

☙ Graph the data from the below tables:

9)

X	1	3	5
y	2	6	10

10)

X	2	6	1
y	3	9	1.5

Ratio and Rates Word Problems

☙ Solve.

1) If the ratio of red to blue marbles is 3:4, find the equivalent ratio when there are 18 red marbles.
2) The recipe uses 2 cups of flour for every 3 cups of sugar. Find the equivalent ratio if you use 8 cups of flour.
3) A car travels 120 miles in 3 hours. At this rate, how far can it travel in 5 hours?
4) The printing press can print 450 pages in 3 hours. At this rate, how many pages can the press print in 8 hours?
5) In a garden, the ratio of rose plants to tulip plants is 7:3. If there are 21 rose plants, how many tulip plants are there?

Converting Between Fractions, Mixed Numbers and Percent

☙ Convert fractions and mixed numbers to percent:

1) $\frac{5}{10}$ 2) $3\frac{7}{25}$ 3) $4\frac{19}{20}$ 4) $\frac{13}{1000}$ 5) $\frac{17}{4}$

☙ Convert percents to fractions:

6) 12%   7) 98%   8) 120%   9) 55%   10) 0.12%

Converting Between Percent and Decimals

☙ Convert.

1) 45%   2) 13%   3) 0.8%   4) 167%   5) 350%
6) 0.12   7) 0.458   8) 1.1   9) 2.36   10) 0.09

Finding Percents of Numbers

Calculate the given percents of each value:

1) 12% *of* 20
2) 30% *of* 14
3) 10% *of* 80
4) 15% *of* 8.2
5) 50% *of* 18
6) 100% *of* 52
7) 120% of 40
8) 10.5% *of* 20
9) 150% *of* 400
10) 300% *of* 100

Discount, Tax and Tip

Find the selling price of each item:

1) Original price of a computer: $420, Tax: 8%, Selling price: $ _____
2) Original price of a chair: $390, Discount: 12%, Selling price: $ _____
3) Food bill: $50, Tip: 10%, Food price: $ _____
4) Original price of a tablet: $80, Discount: 30%, Selling price: $ _____
5) Original price of a house: $630,000, Tax: 1.8%, Selling price: $ _____

Word Problems

Find the answer to each word problem:

1) There are 40 employees in the company. On a certain day, 25 were present. What percentage did you show up for work?
2) A crew is made up of 12 women; the rest are men. If 15% of the crew are women, how many people are in the crew?
3) Maria has $100. She spends 20% on a book. How much money does she spend on the book?
4) The shirt was originally priced at $50. It is on sale for 10% off. What is the sale price of the shirt?
5) A car's value depreciates by 15% each year. If the car is worth $20,000 now, what will it be worth after one year?
6) An item costs $75 before tax. If the sales tax rate is 8%, what is the total cost of the item after tax?
7) A student scores 85% on a test with 40 questions. How many questions did the student answer correctly?
8) The population of 10,000 people increases by 12% each year. What will the population be after one year?
9) A company increases its production by 25% each month. If it produced 400 units last month, how many units would it produce this month?
10) A savings account has an annual interest rate of 5%. If you deposit $1,000, how much interest will you earn in one year?

Answer of Worksheets

Writing and Identifying Ratios

1) $\frac{6}{10} = \frac{3}{5}$
2) $\frac{10}{32} = \frac{5}{16}$
3) $\frac{32}{6} = \frac{16}{3}$
4) $\frac{60}{32} = \frac{15}{8}$
5) $\frac{16}{16} = 1$
6) $\frac{28}{3}$
7) $\frac{24}{10} = \frac{12}{5}$
8) $\frac{3}{24} = \frac{1}{8}$
9) $\frac{38}{24} = \frac{19}{12}$
10) $\frac{27}{28}$

Simplifying Ratios

1) $5:4$
2) $1:4$
3) $3:1$
4) $4:7$
5) $2:7$
6) $14:13$
7) $25:29$
8) $1:3$
9) $47:236$
10) $41:168$

Equivalent Ratios and Rates

1) $2:3 = 4:6 = 6:9 = 8:12$
2) $6:9 = 12:18 = 18:27 = 24:36$
3) $8:7 = 16:14 = 24:21 = 32:28$
4) $14:20 = 7:10 = 21:30 = 28:40$
5) $24:15 = 8:5 = 16:10 = 32:20$
6) 4
7) 24
8) 30
9) 14
10) 18

Proportion in Tables and Graphs

1) Yes 2) No 3) Yes 4) Yes

5)

X	3	9	12
y	10	30	40

6)

X	12	3	30
y	16	4	40

7)

X	35	7	14
y	25	5	10

8)

X	16	1	4
y	24	1.5	6

9)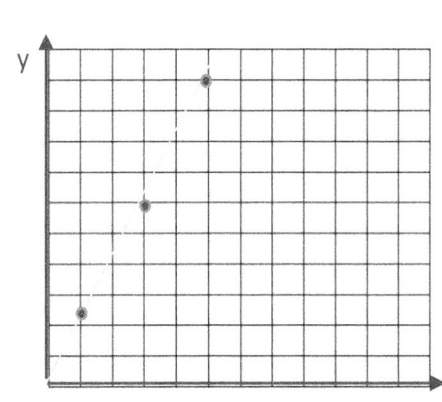

10)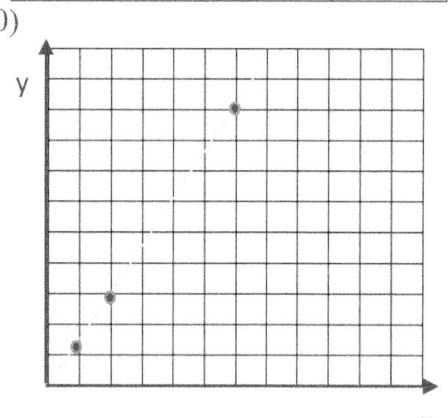

www.mathnotion.com

Ratio and Rates Word Problems

1) 24   2) 12   3) 200   4) 1,200   5) 9

Converting Between Fractions, Mixed Numbers and Percent

1) 50%   2) 328%   3) 495%   4) 1.3%   5) 425%

6) $\frac{12}{100}=\frac{3}{25}$ 7) $\frac{98}{100}=\frac{49}{50}$ 8) $\frac{120}{100}=\frac{6}{5}$ 9) $\frac{55}{100}=\frac{11}{200}$ 10) $\frac{0.12}{100}=\frac{3}{2500}$

Converting Between Percent and Decimals

1) 0.45   2) 0.13   3) 0.008   4) 1.67   5) 3.50

6) 12%   7) 45.8%   8) 110%   9) 236%   10) 9%

Finding Percents of Numbers

1) 2.4
2) 4.2
3) 8
4) 1.23
5) 9
6) 52
7) 48
8) 2.1
9) 600
10) 300

Discount, Tax and Tip

1) $435.60
2) $343.20
3) $45
4) $56
5) $ 641,340

Word Problems

1) 62.5%
2) 80 people
3) $20
4) $45
5) $17,000
6) $81
7) 34 questions
8) 11,200
9) 500 units
10) $50

Chapter 7: Exponents and Radicals Expression

Topics that you will learn in this chapter:

- Evaluating Powers
- Finding Missing Exponents
- Adding and Subtracting Exponents
- Multiplication Property of Exponents
- Division Property of Exponents
- Powers of Products and Quotients
- Negative Exponents and Negative Bases
- Decimal and Fractional Bases
- Powers of Ten
- Scientific Notation
- Square Roots
- Word Problems
- Worksheets
- Answer of Worksheets

Evaluating Powers

Evaluating powers involves understanding and applying the concept of exponents. Here's a straightforward guide:

1. **Identify the Base and Exponent**:
   - The base is the number being multiplied.
   - The exponent tells you how many times to multiply the base by itself.
   - Mathematical symbol: a^x (a is the base, x is the exponent)

2. **Multiply the Base by Itself**:
   - Multiply the base by itself as many times as indicated by the exponent.

   $$a^x = \underbrace{a \times a \times \ldots \times a}_{X \text{ times}}$$

3. **Calculate the Result**:
   - Perform the multiplication.

Examples:

1) Evaluate 3^5:

   Solution:
   1. Identify the Base and Exponent: The base is 3 and the exponent is 5.
   2. Multiply the Base by Itself: $3^5 = 3 \times 3 \times 3 \times 3 \times 3$
   3. Answer: $3^5 = 243$.

2) Fill the blanks with the correct number:

   a) $4^{\square} = 4 \times 4 \times 4$ b) $\square^6 = 2 \times 2 \times 2 \times 2 \times 2 \times 2$

   Solution:

   a) The base is 4 and the number of times that the base is multiplied by itself is 3, so the exponent is 3: $4^{\boxed{3}} = 4 \times 4 \times 4$.

   b) The exponent is 6 and the number 2 is multiplied by itself six times, so the base is 2: $\boxed{2}^6 = 2 \times 2 \times 2 \times 2 \times 2 \times 2$.

3) Evaluate $(2^2)^3$:

   Solution:
   1. Identify the Base and Exponent: The base is 2^2 (or $2 \times 2 = 4$) and the exponent is 3.
   2. Multiply the Base by Itself: $(2^2)^3 = 2^2 \times 2^2 \times 2^2 = 4 \times 4 \times 4$.
   3. Answer: $(2^2)^3 = 64$.

www.mathnotion.com

Finding Missing Exponents

Finding a missing exponent can be simplified with step-by-step approaches. Here's a method they can use: Use **guess and check** or **prime factorization** if you're not sure right away.

If we have $a^x = b$, do following steps:

1. **Understand the Problem**:
   - We know that a raised to some power x equals b.
2. **Think of What Number Multiplied by Itself Equals b (guess and check)**:
   - Start with small numbers and multiply until you reach b.
3. We can also use the **Prime Factorization** method in chapter 4 too (Start with the smallest prime number, divide the number by the prime, repeat the process and finally list all prime factors)
4. **Count the Number of Times**.
5. **Write the Exponent**.

Examples:

1) Find the missing exponent x in $3^x = 27$.

   Solution:
   1. **Understand the Problem**: We know that 3 raised to some power x equals 27.
   2. **Think of What Number Multiplied by Itself Equals 27:** Start with small numbers and multiply until you reach 27: $3 \times 3 = 9$ and $3 \times 3 \times 3 = 27$
   3. **Count the Number of Times**: Notice that we multiplied 3 by itself 3 times to get 27.
   4. **Write the Exponent:** So, $3^3 = 27$.

2) Find the missing exponent x in $2^x = 128$.

   Solution:
   1. **Understand the Problem**: We know that 2 raised to some power x equals 128.
   2. **Think of What Number Multiplied by Itself Equals 128 (or use the Prime Factorization method):** $128 = 2 \times 2 \times 2 \times 2 \times 2 \times 2 \times 2$.
   3. **Count the Number of Times**: Notice that we multiplied 2 by itself 7 times to get 128.
   4. **Write the Exponent:** So, $2^7 = 128$.

www.mathnotion.com

Adding and Subtracting Exponents

Before performing any addition or subtraction between exponents we need to observe whether the base and exponents are the same or not:

Same Bases and Exponents:

1. Arrange the similar base and exponent terms together.
2. Now perform addition or subtraction as per need between the base of terms.

Different Bases and Exponents: Solve them individually.

☑ In general, we calculate each item and then add/subtract the result.

Examples:

1) Solve $6^3 + 6^3$.

   Solution:

   Here the base and exponents are the same. So, we are solving them together:

   $6^3 + 6^3 = 2\,(6^3) = 2 \times 6 \times 6 \times 6 = 432$

2) Solve $9^4 - 7^3$.

   Solution:

   Here the base and the exponents are different. So, we are solving them individually:

   $9^4 - 7^3 = 6{,}561 - 343 = 6{,}218.$

3) Solve $2^3 + 3^4 - 5^2 + 4^3$:

   Solution:

   All the bases and exponents are different so we can use the general method and calculate each item individually:

   $2^3 = 8 \quad 3^4 = 81 \quad 5^2 = 25 \quad 4^3 = 64$

   Now add and subtract the results: $2^3 + 3^4 - 5^2 + 4^3 = 8 + 81 - 25 + 64 = 128.$

4) Solve $5^3 + 2^4 + 5^3 + 2^4 + 5^3$:

   Solution:

   1. Arrange the similar base and exponent terms together: $5^3 + 5^3 + 5^3 + 2^4 + 2^4$
   2. Now perform addition or subtraction:

      $5^3 + 5^3 + 5^3 = 3\,(5^3) = 3 \times 125 = 375$

      $2^4 + 2^4 = 2\,(2^4) = 2 \times 16 = 32$

   The result: $375 + 32 = 407$

Multiplication Property of Exponents

Multiplying exponents using the properties of exponents can simplify calculations and expressions. Here's a guide to how to do it:

Product of Power Property:

Rule: $a^n \times a^m = a^{n+m}$: When multiplying two powers with the same base, add the exponents.

Power of Power Property:

Rule: $(a^n)^m = a^{n \times m}$: When raising power to another exponent, multiply the exponents.

Power of a Product Property:

Rule: $(ab)^n = a^n \times b^n$: When raising a product to power, apply the exponent to each factor.

Examples:

1) Evaluate $3^2 \times 3^4$ using the product of powers property:

   Solution:
   1. Identify the bases and exponents: Both terms have the same base (3) and exponents are 2 and 4.
   2. Apply the product of power property: When multiplying two powers with the same base, add the exponents: $3^2 \times 3^4 = 3^6$.
   3. Calculate the result: $3^6 = 3 \times 3 \times 3 \times 3 \times 3 \times 3 = 729$.

2) Evaluate $(5^2)^3$ using the power of a power property:

   Solution:
   1. Identify the bases and exponents: The base is 5 and the exponents are 2 and 3.
   2. Apply the power of power property: When raising a power to another exponent, multiply the exponents: $(5^2)^3 = 5^6$
   3. Calculate the result: $5^6 = 5 \times 5 \times 5 \times 5 \times 5 \times 5 = 15{,}625$.

3) Evaluate $(7 \times 2)^3$ using the power of a product property:

   Solution:
   1. Identify the bases and exponents: The base is 7×2 and the exponents is 3.
   2. Apply the power of product property: When raising a product to a power, distribute the exponent to each factor: $(7 \times 2)^3 = 7^3 \times 2^3$
   3. Calculate the result: $7^3 = 343$ and $2^3 = 8$ so, $7^3 \times 2^3 = 343 \times 8 = 2{,}744$.

Division Property of Exponents

The Division Property of Exponents helps simplify expressions where you divide powers with the same base. Here are the key properties:

Quotient of Powers Property:

Rule: $a^n \div a^m = \frac{a^n}{a^m} = a^{n-m}$: When dividing two powers with the same base, subtract the exponents.

Power of Quotient Property:

Rule: $(a \div b)^n = \left(\frac{a}{b}\right)^n = \frac{a^n}{b^n}$: When raising a quotient to a power, apply the exponent to both the numerator and the denominator.

Zero Exponent Property:

Rule: $a^0 = 1$ (where $a \neq 0$): Any non-zero number raised to the power of 0 is 1.

Examples:

1) Evaluate $\frac{7^8}{7^5}$ using quotient of powers property:

   Solution:
   1. Identify the bases and exponents: Both terms have the same base (7) and exponents are 8 and 5.
   2. Apply the quotient of power property: When dividing two powers with the same base, subtract the exponents: $\frac{7^8}{7^5} = 7^{8-5} = 7^3$
   3. Calculate the result: $7^3 = 7 \times 7 \times 7 = 343$.

2) Evaluate $\left(\frac{9}{10}\right)^2$ using power of quotient property:

   Solution:
   1. Identify the bases and exponents: The base is $\frac{9}{10}$ and the exponents is 2.
   2. Apply the power of quotients property: When raising a quotient to a power, apply the exponent to both the numerator and the denominator.: $\left(\frac{9}{10}\right)^2 = \frac{9^2}{10^2}$
   3. Calculate the result: $\frac{9^2}{10^2} = \frac{9 \times 9}{10 \times 10} = \frac{81}{100}$.

3) Evaluate $\left(\frac{4}{7}\right)^0 \div 5^3$:

   Solution: By using zero exponent property: $\left(\frac{4}{7}\right)^0 = 1$ and $1 \div 5^3 = \frac{1}{5^3} = \frac{1}{5 \times 5 \times 5} = \frac{1}{125}$.

www.mathnotion.com

Negative Exponents and Negative Bases

Negative Exponents:

- **Definition:** A negative exponent indicates that the base should be taken to reciprocal power.
- $a^{-n} = \frac{1}{a^n}$.

Negative Bases:

- **Definition:** When a base is negative, the sign of the result depends on whether the exponent is even or odd.
- $(-a)^n$.

Examples:

1) Rewrite the exponential numbers as positive exponents and then calculate them.

   a. 4^{-3} b. 2^{-5} c. 6^{-1}

   Solution:

   Using negative exponents rule:

   a. $4^{-3} = \frac{1}{4^3} = \frac{1}{4 \times 4 \times 4} = \frac{1}{64}$

   b. $2^{-5} = \frac{1}{2^5} = \frac{1}{2 \times 2 \times 2 \times 2 \times 2} = \frac{1}{32}$

   c. $6^{-1} = \frac{1}{6}$

2) Evaluate $\left(-\frac{2}{3}\right)^4 \times \frac{2^{-2}}{3^{-6}}$ using power rules:

   Solution:

   For the left item, the base is negative, but the exponent is 4 so: $\left(-\frac{2}{3}\right)^4 = \left(\frac{2}{3}\right)^4$.

   1. Using the power of quotient property: $\left(\frac{2}{3}\right)^4 = \frac{2^4}{3^4}$.
   2. For the right item, the exponents in numerator and denominator are negative so: $2^{-2} = \frac{1}{2^2}$ and $3^{-6} = \frac{1}{3^6}$ and then $\frac{2^{-2}}{3^{-6}} = \frac{\frac{1}{2^2}}{\frac{1}{3^6}} = \frac{1}{2^2} \div \frac{1}{3^6} = \frac{1}{2^2} \times \frac{3^6}{1} = \frac{3^6}{2^2}$.
   3. By using quotient of powers property, product of power property and arranging numbers with the same bases we have: $\left(-\frac{2}{3}\right)^4 \times \frac{2^{-2}}{3^{-6}} = \frac{2^4}{3^4} \times \frac{3^6}{2^2} = \frac{2^4}{2^2} \times \frac{3^6}{3^4} = 2^2 \times 3^2 = 6^2$
   4. The final answer: $6^2 = 6 \times 6 = 36$.

Decimal and Fractional Bases

When the bases are decimas or fractions, you handle them similarly to whole numbers, pay attention to the key points:

- **Precision:** Be careful with multiplication or division and ensure accuracy, especially with decimals.
- **Same Rules Apply:** Exponential properties hold true regardless of whether the base is a whole number, decimal, or fraction.
  - ☑ For negative powers of fractions, you can use this method: $(\frac{a}{b})^{-n} = (\frac{b}{a})^n$
- **Simplifying:** Sometimes converting decimals to fractions and simplifying them, make your work easier.

Examples:

1) Evaluate $(\frac{2}{5})^3$ and 0.1^{-2}

   Solution:
   1. Identify the Base and Exponent: For first item, the base is $\frac{2}{5}$ and the exponent is 3 and for the second item the base is 0.1 and the exponent is -2.
   2. Based on positive and negative power rules: $(\frac{2}{5})^3 = \frac{2}{5} \times \frac{2}{5} \times \frac{2}{5} = \frac{8}{125}$ and $0.1^{-2} = \frac{1}{0.1^2} = \frac{1}{0.1 \times 0.1} = \frac{1}{0.01} = \frac{1}{\frac{1}{100}} = 1 \div \frac{1}{100} = 1 \times \frac{100}{1} = 100$.

2) Evaluate $0.8^3 \times \left(\frac{5}{4}\right)^{-2} \times 1.2^5$:

   Solution:

   At first glance, none of the bases and exponents are the same, but if we convert 0.8 to fraction and simplify that, we have: $0.8 = \frac{8}{10} = \frac{4}{5}$. Therefore, according to negative exponent, power of quotients and quotients of powers properties: $\left(\frac{5}{4}\right)^{-2} = \left(\frac{4}{5}\right)^2$. Now we have:

   $0.8^3 \times \left(\frac{5}{4}\right)^{-2} = (\frac{4}{5})^3 \times (\frac{4}{5})^2 = (\frac{4}{5})^5$ (basing product of powers property). Finally, by using the power of product property: $(\frac{4}{5})^5 \times 1.2^5 = (\frac{4}{5} \times 1.2)^5 = (0.8 \times 1.2)^5$.

   The final answer: $0.8^3 \times \left(\frac{5}{4}\right)^{-2} \times 1.2^5 = 0.96^5$.

Powers of Ten

Powers of Ten are fundamental in understanding place value, scientific notation, and simplifying calculations. Here's a comprehensive and short look:

- **Basic Property**: 10^n for multiplying 10 by itself n times.
- **Zero Exponent**: $10^0 = 1$.
- **Negative Exponent**: $10^{-n} = \frac{1}{10^n}$
- **Multiplying**: Add the exponents: $10^n \times 10^m = 10^{n+m}$
- **Dividing**: Subtract the exponents: $10^n \div 10^m = 10^{n-m}$
- **Power of a Power**: Multiply the exponents. $(10^n)^m = 10^{n \times m}$
- **Power of a Product**: Apply the exponent to each factor:

  $(a \times 10^n) \times (b \times 10^m) = a \times b \times 10^{n+m}$

- **Power of a Quotient**: Apply the exponent to both the numerator and the denominator:

  $\frac{a \times 10^n}{b \times 10^m} = \frac{a}{b} \times 10^{n-m}$

Example:

Evaluate $(10^2 \times 10^{-3})^4$ using properties of exponents:

Solution:

1. **Apply the Power of a Product Property:** Distribute the exponent 4 to each factor inside the parentheses:

   $$(10^2 \times 10^{-3})^4 = (10^2)^4 \times (10^{-3})^4$$

2. **Apply the Power of a Power Property**: Multiply the exponents:

   $(10^2)^4 = 10^{2 \times 4} = 10^8$ and $(10^{-3})^4 = 10^{-3 \times 4} = 10^{-12}$

3. **Multiply the Powers of Ten:** Use the product of powers property:

   $10^8 \times 10^{-12} = 10^{8-12} = 10^{-4}$

4. **Final Answer:**

   $(10^2 \times 10^{-3})^4 = 10^{-4} = \frac{1}{10^4} = \frac{1}{10 \times 10 \times 10 \times 10} = \frac{1}{10000}$.

Scientific Notation

Scientific notation is a way of expressing very large or very small numbers in a compact form. It uses the format: $a \times 10^n$ where:

- a is a number greater than or equal to 1 and less than 10.
- n is an integer (positive or negative).

Steps to Convert a Number to Scientific Notation:

1. **Identify Significant Figures**: Move the decimal point to create a number between 1 and 10.
2. **Determine the Exponent:** Count places the decimal moves. Left moves are positive, right moves are negative.
3. **Combine:** Form the expression $a \times 10^n$.

Examples:

1) Convert 45,000 to scientific notation:
   Solution:
   1. Identify the Significant Figures: The significant figures are 4.5 (ignore trailing zeros for now).
   2. Determine the Exponent:
      - Count how many places the decimal point moves to the left to position it after the first digit.
      - In 45,000, the decimal moves 4 places: 45000⇒4.5000
   3. Write in Scientific Notation: Combine the significant figures and the power of ten: 4.5×10^4.

2) Convert 0.0032 to scientific notation:
   Solution:
   1. Identify the Significant Figures: The significant figures are 3.2.
   2. Determine the Exponent:
      - Count how many places the decimal point moves to the right to position it after the first digit.
      - In 0.0032, the decimal moves 3 places: 0.0032⇒3.2.
   3. Write in Scientific Notation: Combine the significant figures and the power of ten: 3.2×10^{-3}.

Square Roots

Finding square roots involves identifying the number that, when multiplied by itself, equals the given number. Here's a step-by-step method:

Steps to Find Square Roots:

1. **Perfect Squares:** Recognize if the number is a perfect square (e.g., 1, 4, 9, 16, 25, etc.).
2. **Prime Factorization**: Break down the number into its prime factors and pair them up.
3. **Using Estimation:** Estimate the square root if it's not a perfect square.

Examples:

1) Find the square root of 36:

   Solution:
   1. Prime factorization: $36 = 2 \times 2 \times 3 \times 3$
   2. Pairs: (2×2) and (3×3)
   3. Take one number from each pair: $\sqrt{36} = 2 \times 3 = 6$.

2) Find the square root of $\frac{0.01}{25}$.

   Solution:
   1. First, we try to simplify the fraction: $\frac{0.01}{25} = \frac{\frac{1}{100}}{25} = \frac{1}{100} \div 25 = \frac{1}{100} \times \frac{1}{25} = \frac{1}{2500}$
   2. Prime Factorization of 2,500: $2{,}500 = 2 \times 2 \times 5 \times 5 \times 5 \times 5$
   3. Pairs: (2×2) and (5×5) and (5×5)
   4. Take one number from each pair: $\sqrt{\frac{1}{2500}} = \frac{1}{\sqrt{2500}} = \frac{1}{5 \times 5 \times 2} = \frac{1}{50}$.

3) Estimate $\sqrt{20}$.

   Solution:
   - $\sqrt{20}$ is between 4 (since $4^2=16$) and 5 (since $5^2=25$).
   - Try $4.5 \times 4.5 = 20.25$.
   - Adjust accordingly, so $\sqrt{20} \approx 4.5$.

Word problems

Like previous chapters and here for solving word problems involving exponents and radicals we should follow these steps:

1. **Understand the Problem:** Read the problem carefully to identify the quantities involved and then determine what the problem is asking for.
2. **Identify the Exponential Relationship**: Look for phrases like "doubles," "triples," "squares," etc., which indicate an exponential relationship, then recognize the base and the exponent.
3. **Set Up the Equation**: Translate the word problem into a mathematical equation involving exponents.
4. **Apply Exponent and Radical Rules**: Use appropriate exponent and radical rules (multiplication, division, power of power, square roots) to simplify the equation and then calculate and solve them.
5. **Verify the Solution**: Check if the solution makes sense in the context of the problem.

Example:

If you deposit $50 into a savings account that doubles in value every year, how much will the account be worth after 5 years?

Solution:

1. Identify the Growth Pattern:
   - The money doubles every year.
   - This can be expressed using an exponent.

2. Set Up the Equation:
   - Initial deposit: $50
   - Value after 5 years: 50×2^5

3. Calculating the Exponent: $2^5 = 2 \times 2 \times 2 \times 2 \times 2 = 32$
4. Find the Total Value: $50 \times 2^5 = 50 \times 32 = 1,600$, so the account will worth $1,600 after 5 years.

Worksheets

Evaluating Powers and Finding Missing Exponents

Evaluate.

1) 4^2
2) 2^7
3) 9^3
4) $(\frac{2}{5})^4$
5) 0.03^2

Fill the blanks with the correct number:

1) $6^{\square} = 216$
2) $5^{\square} = 125$
3) $\square^2 = 121$
4) $\square^3 = \frac{1}{27}$
5) $0.1^{\square} = 0.0001$

Adding and Subtracting Exponents

Add and subtract.

1) $2^2 + 3^2$
2) $5^3 - 2^5$
3) $5^4 + 5^4 + 5^4$
4) $3^3 - 3^2 + 3^4$
5) $8^2 + 2^6 + 8^2 + 2^6$
6) $a^3 - b^2 - a^3$
7) $(\frac{1}{4})^2 + (\frac{1}{2})^3 - (\frac{2}{3})^2 - (\frac{1}{3})^3$
8) $5^2 - 3^3 + 4^2$
9) $10^2 + 10^3 + 10^4 + 10^5$
10) $0^5 + 9^2 + 8^3 - 3^4 - 2^9$

Multiplication Property of Exponents

Multiply the expressions and write the result of the expression as an exponential number:

1) $4^5 \times 4^3$
2) $2^4 \times 2^3 \times 2^5$
3) $7^5 \times 7 \times 7^3 \times 7^2$
4) $(9^2)^5$
5) $25^4 \times 2^4$
6) $(3^2 \times 5^2 \times 2^2)^3$
7) $(3^2 \times 3^2 \times 3^2)^2$
8) $5^3 \times 7^3 \times 35^2$
9) $12^5 \times 6^2 \times 2^2 \times 9^7$
10) $(9^3 \times 5^{11} \times 3^5 \times 2^{11})^{10}$

Division Property of Exponents

Divide the expressions and write the result of the expression as an exponential number:

1) $6^7 \div 6^2$
6) $\frac{11^5 \times 11^2}{10^3 \times 10^4}$

2) $9^{10} \div 9$

3) $\frac{5^4}{5^3}$

4) $\frac{3^{11}}{3^5}$

5) $\frac{7^5 \times 7^2}{7^3}$

7) $\frac{9^5 \div 9^4}{9}$

8) $\frac{2^7 \times 3^7 \times 6^5}{24^{12} \div 4^{12}}$

9) $\frac{15^4 \times (2^3)^2 \times 5^2 \times 3^2}{5^3 \times 6^3}$

10) $((10^3 \div 5^3) \times (14^5 \div 7^5))^{10}$

Negative Exponents and Negative Bases

✍ Evaluate the expressions and write the result as an exponential number:

1) $2^{-2} \times 2^3$

2) $5^4 \div 5^{-1}$

3) $-3^5 \times 3^{-4} \times 3^{-2}$

4) $(11^5 \div 11^{-3})^2$

5) $18^{-4} \times (-2)^{-4}$

6) $\frac{4^{-2} \times 2^{-2}}{(-8)^3}$

7) $\left(-\frac{2}{3}\right)^{-2} \times \left(\frac{4}{6}\right)^5$

8) $(-(18^{-2} \div 6^{-2}) \times (15 \div 5)^2)^{-4}$

9) $\frac{2^{-6}}{3^4} \times \frac{3^{-5}}{2^3}$

10) $\left(\frac{7^{-2}}{7^2} \div \frac{7^2}{7^{-6}}\right) \times \left(\frac{3^{-2}}{3^4} \div \frac{3^7}{3}\right)$

Decimal and Fractional Bases

✍ Evaluate the expressions and write the result as an exponential number:

1) $\left(\frac{4}{5}\right)^{-2}$

2) 0.02^3

3) $\left(\frac{3}{5}\right)^5 \times \left(\frac{6}{10}\right)^4$

4) $(0.5^5 \div 0.5^2)^{-1}$

5) $\left(\frac{18}{9}\right)^2 \times \frac{(-2)^4}{2} \times \frac{30}{15}$

6) $1.2^4 \times \left(\frac{12}{10}\right)^{-3} \times \left(\frac{6}{5}\right)^{-2}$

7) $(0.001)^{-2} \times (0.001)^2$

8) $(0.5^2) \times \left(\frac{1}{2}\right)^5 \times \left(\frac{4}{8}\right)^{-3}$

9) $\frac{2^{-6}}{3^{-6}} \div \frac{6^{-5}}{9^{-5}} \times \frac{4^2}{6^2}$

10) $(0.6^2 \div 3^2)^{-3}$

Powers of Ten

✍ Evaluate powers of ten:

1) 10^4

2) $10^{-3} \times 10^3$

3) $10^2 \times 10^5$

4) $10^7 \div 10^5 \div 10^2$

5) $1{,}000^5 \times 100^2 \times 100^{-6}$

6) $(10^{-1})^3$

7) $(-3 \times 10^4) \times (5 \times 10^{-2})$

8) $(-100)^2 \times 0.01^2$

9) $\frac{28 \times 10^6}{4 \times 10^{-8}}$

10) $10^{-5} \div 100^5 \times 0.001^{-5}$

Scientific Notation

🖎 Convert numbers below to scientific notation:

1) 0.00036
2) 0.12589
3) 7,582.36
4) 800,000,000
5) 0.0018002
6) 480,000,000,000
7) 0.0000004
8) 5,614.2311
9) 12.00008
10) 46,500,000

Square Roots

🖎 Find the square root of the numbers:

1) 49
2) 100
3) 3,600
4) 810,000
5) 0.16
6) 0.0004
7) 17
8) 31
9) 99
10) 125

Word problems

🖎 Find the answer to each word problem:

1) A bacteria culture doubles every hour. If you start with 2 bacteria, how many bacteria will there be after 6 hours?
2) A small town's population is 1,000 people. Each year, the population triples. What will the population be like after 3 years?
3) A cube has a side length of 2 cm. What is the volume of the cube?
4) A square garden has an area of 49 square meters. What is the length of one side of the garden?
5) You have a piece of paper that is 0.01 inches thick. If you fold the paper in half 10 times, how thick will the stack be?
6) If the volume of a cube is 64 cubic centimeters, what is the length of one side of the cube?
7) In a video game, you earn double the points at every level. If you start with 5 points, how many points will you have after 8 levels?
8) A scientist has a sample of a radioactive substance that halves in quantity every year. If she starts with 128 grams, how much will be left after 7 years?
9) A rectangle's width is $\sqrt{49}$ meters, and its length is 2^3 meters. What is the area of the rectangle?

Answer of Worksheets

Evaluating Powers and Finding Missing Exponents

1) 16
2) 128
3) 729
4) $\frac{16}{625}$
5) 0.0009
6) 3
7) 3
8) 11
9) $\frac{1}{3}$
10) 4

Adding and Subtracting Exponents

1) 13
2) 93
3) 1,875
4) 99
5) 256
6) $-b^2$
7) $\frac{-127}{432}$
8) 14
9) 111,100
10) 0

Multiplication Property of Exponents

1) 4^8
2) 2^{12}
3) 7^{11}
4) 9^{10}
5) 50^4
6) 30^6
7) 3^{12}
8) 35^5
9) 108^7
10) 30^{110}

Division Property of Exponents

1) 6^5
2) 9^9
3) 5^1
4) 3^6
5) 7^4
6) $(\frac{11}{10})^7$
7) 1
8) 1
9) 30^3
10) 2^{80}

Negative Exponents and Negative Bases

1) 2
2) 5^5
3) $\frac{-1}{3}$
4) 11^{16}
5) $\frac{1}{36^4}$
6) $\frac{-1}{8^5}$
7) $(\frac{2}{3})^3$
8) 1
9) $\frac{1}{6^9}$
10) $\frac{1}{21^{12}}$

Decimal and Fractional Bases

1) $\left(\frac{5}{4}\right)^2$
2) 0.000008
3) $\left(\frac{3}{5}\right)^9$
4) 2^3
5) 2^6
6) $\frac{5}{6}$
7) 1
8) 0.5^4
9) $\frac{2}{3}$
10) 5^6

Powers of Ten

1) 10,000
2) 1
3) 10,000,000
4) 1
5) 10,000,000
6) 0.001
7) -1,500
8) 1
9) 700,000,000,000,000
10) 1

Scientific Notation

1) 3.6×10^{-4}
2) 1.2589×10^{-1}
3) 7.58236×10^3
4) 8×10^8
5) 1.8002×10^{-3}
6) 4.8×10^{11}
7) 4×10^{-7}
8) 5.6142311×10^3
9) 1.200008×10^1
10) 4.65×10^7

Square Roots

1) 7
2) 10
3) 60
4) 900
5) 0.4
6) 0.02
7) ≈ 4.1
8) ≈ 5.6
9) ≈ 9.9
10) ≈ 11.2

Word problems

1) 128 bacteria
2) 27,000 people
3) 8 cubic centimeters
4) 7 meters
5) 10.24 inches
6) 4 centimeters
7) 1,280 points
8) 1 gram
9) 56 square meters

www.mathnotion.com

Reference Measurement

LENGTH	
Customary	**Metric**
1 mile (mi) = 1,760 yards (yd)	1 kilometer (km) = 1,000 meters (m)
1 yard (yd) = 3 feet (ft)	1 meter (m) = 100 centimeters (cm)
1 foot (ft) = 12 inches (in.)	1 centimeter(cm) = 10 millimeters(mm)
VOLUME AND CAPACITY	
Customary	**Metric**
1 gallon (gal) = 4 quarts (qt)	1 liter (L) = 1,000 milliliters (mL)
1 quart (qt) = 2 pints (pt.)	
1 pint (pt.) = 2 cups ©	
1 cup © = 8 fluid ounces (Fl oz)	
WEIGHT AND MASS	
Customary	**Metric**
1 ton (T) = 2,000 pounds (lb.)	1 kilogram (kg) = 1,000 grams (g)
1 pound (lb.) = 16 ounces (oz)	1 gram (g) = 1,000 milligrams (mg)
Time	
1 year = 12 months	
1 year = 52 weeks	
1 week = 7 days	
1 day = 24 hours	
1 hour = 60 minutes	
1 minute = 60 seconds	

Chapter 8: Measurements

Topics that you will learn in this chapter:
- Reference Measurement
- Metric Length Measurement
- Customary Length Measurement
- Metric Capacity Measurement
- Customary Capacity Measurement
- Metric Weight and Mass Measurement
- Customary Weight and Mass Measurement
- Temperature
- Time
- Word Problems
- Worksheets

Reference Measurement

Reference measurement refers to a standard or benchmark used to compare and evaluate other measurements. It's like having a base point to ensure consistency and accuracy in various fields, such as science, engineering, and daily life.

General Steps for Converting Units:

1. **Identify the Units:** Clearly identify the unit you want to convert from (the source unit) and the unit you want to convert to (the target unit)
2. **Find the Conversion Factor:** Determine the conversion factor between the two units. This is the ratio that relates to the two units.
3. **Set Up the Conversion:**
   - Write the original measurement with its unit
   - Multiply this by the conversion factor, making sure to arrange it so that the unwanted unit cancels out.
4. **Perform the Calculation:**
   - Converting Smaller to Larger Units: When converting from a smaller unit to a larger one, you divide.
   - Converting Larger to Smaller Units: When converting from a larger unit to a smaller one, you multiply
5. **Check the Units:** Ensure that the units in the final answer are the desired target units.
   - ☑ Use a Conversion Table or Online Converter: These tools can provide conversion factors for various units.
   - ☑ Practice Regularly: The more you practice, the better you'll become at unit conversions.

Example:

Convert 10 inches to centimeters.

Solution:

1. Identify Units: Source unit: inches and Target unit: centimeters.
2. Find Conversion Factor: 1 inch = 2.54 centimeters.
3. Set Up Conversion: 10 inches × (2.54 centimeters / 1 inch)
4. Perform Calculation: When converting from a larger unit to a smaller one, we must multiply, so: 10 × 2.54 = 25.4 Therefore, 10 inches is equal to 25.4 centimeters.

Metric Length Measurement

Converting metric length measurements involves shifting between different units within the metric system, such as millimeters (mm), centimeters (cm), meters (m), and kilometers (km). Here's a clear guide:

Conversion Steps:

1. **Understand the Metric Prefixes:**
    a. 1 kilometer (km) = 1,000 meters (m)
    b. 1 meter (m) = 100 centimeters (cm)
    c. 1 centimeter (cm) = 10 millimeters (mm)
2. **Determine the Conversion Factor:** Identify the units you are converting from and to.
3. **Set Up the Conversion:** Use the conversion factor to multiply or divide the given value.

Examples:

1) Convert 0.6 meters to centimeters:

    Solution:
    1. Understand the Metric Prefixes: 1 meter = 100 centimeters.
    2. Determine the Conversion Factor: Converting meters to centimeters involves multiplying by 100.
    3. Set Up the Conversion: 0.6 meter × 100 = 60 centimeters.

2) Convert 150 millimeters to meters:

    Solution:
    1. Understand the Metric Prefixes: 1 meter = 100 centimeters = 100 ×10 millimeters = 1,000 millimeters.
    2. Determine the Conversion Factor: Converting millimeters to meters involves dividing by 1,000.
    3. Set Up the Conversion: 150 millimeters ÷ 1,000 = 0.150 meters.

3) Convert 14 kilometers to meters:

    Solution:
    1. Understand the Metric Prefixes: 1 kilometer = 1,000 meters.
    2. Determine the Conversion Factor: Converting kilometers to meters involves multiplying by 1,000.
    3. Set Up the Conversion: 41 kilometers × 1,000= 41,000 meters.

www.mathnotion.com

Customary Length Measurement

The customary system of measurement includes these units of length:

Conversion Steps:

1. **Understand the Metric Prefixes:**
   a. Inch (in): The smallest unit, often used for precise measurement.
   b. Foot (ft): Equal to 12 inches.
   c. Yard (yd): Equal to 3 feet or 36 inches.
   d. Mile (mi): Equal to 5,280 feet or 1,760 yards.
2. **Determine the Conversion Factor:** Identify the units you are converting from and to.
3. **Set Up the Conversion:** Use the conversion factor to multiply or divide the given value.

Examples:

1) Convert 5 yards to feet:

   Solution:
   1. Understand the Metric Prefixes: 1 yard = 3 feet
   2. Determine the Conversion Factor: Converting yards to feet involves multiplying by 3.
   3. Set Up the Conversion: 5 yards × 3 = 15 feet.

2) Convert 12 inches to yard:

   Solution:
   1. Understand the Metric Prefixes: 1 yard = 3 feet = 3 ×12 inches = 36 inches.
   2. Determine the Conversion Factor: Converting inches to yards involves dividing by 36.
   3. Set Up the Conversion: 12 inches ÷ 36 ≈ 0.3 yard.

4) Convert 26.2 miles to inches:

   Solution:
   1. Understand the Metric Prefixes: 1 miles= 5,280 feet = 5,280 × 12inches = 63,360 inches.
   2. Determine the Conversion Factor: Converting miles to inches involves multiplying by 63,360.
   3. Set Up the Conversion: 26.2 miles × 63,360= 1, 660,032 inches.

www.mathnotion.com

Metric Capacity Measurement

Metric capacity measurement refers to the system used to measure the volume or capacity of liquids and solids in the metric system. Here's a detailed overview:

Conversion Steps:

1. **Understand the Metric Prefixes:**
    a. 1 kiloliter (KL) = 1,000 liters (L)
    b. 1 liter (L) = 1,000 milliliters (mL)
    c. 1 liter (L) = 100 centiliters (CL)
    d. 1 liter (L) = 10 deciliters (dL)
    e. 1 milliliter (mL) = 0.001 liters (L)
    f. 1 centiliter (c L) = 0.01 liters (L)
    g. 1 deciliter (dL) = 0. 1 liter (L)

- CC stands for cubic centimeters which is used in medicine industries. In metric system 1CC = 1 milliliter (mL) and they are used interchangeably.

2. **Determine the Conversion Factor:** Identify the units you are converting from and to.
3. **Set Up the Conversion:** Use the conversion factor to multiply or divide the given value.

Examples:

1) Convert 2,500 milliliters to liters:

   Solution:
   1. Determine the Starting Unit and Desired Unit:
      - Starting unit: milliliters (mL)
      - Desired unit: liters (L)
   2. Determine the Conversion Factor: Converting milliliters to liters involves dividing by 1,000.
   3. Set Up the Conversion: 2,500 milliliters ÷ 1,000 = 2.5 liters

2) Convert 0.75 kiloliters to liters:

   Solution:
   1. Determine the Starting Unit and Desired Unit:
      - Starting unit: kiloliters (KL)
      - Desired unit: liters (L)
   2. Determine the Conversion Factor: Converting kiloliters to liters involves multiplying by 1,000.
   3. Set Up the Conversion: 0.75 kiloliters × 1,000 = 750 liters.

www.mathnotion.com

Customary Capacity Measurement

Customary capacity measurement is a system used in the United States to measure the volume or capacity of liquids and dry substances. This system includes units such as teaspoons, tablespoons, fluid ounces, cups, pints, quarts, and gallons.

Conversion Steps:

1. **Understand the Metric Prefixes:**
   a. 1 tablespoon (tbsp) = 3 teaspoons (tsp)
   b. 1 fluid ounce (fl oz) = 2 tablespoons (tbsp)
   c. 1 cup (c) = 8 fluid ounces (fl oz)
   d. 1 pint (pt) = 2 cups (c) 1 quart (qt) = 2 pints (pt)
   e. 1 gallon (gal) = 4 quarts (qt)
2. **Determine the Conversion Factor:** Identify the units you are converting from and to.
3. **Set Up the Conversion:** Use the conversion factor to multiply or divide the given value.

Examples:

1) Convert 4 cups to fluid ounces:
   Solution:
   1. Determine the Starting Unit and Desired Unit:
      - Starting unit: cups (c)
      - Desired unit: fluid ounces (fl oz)
   2. Determine the Conversion Factor: Converting cups to fluid ounces involves multiplying by 8.
   3. Set Up the Conversion: 4 cups × 8 = 32 fluid ounces

2) Convert 48 quarts to gallons:
   Solution:
   1. Determine the Starting Unit and Desired Unit:
      - Starting unit: quarts (qt)
      - Desired unit: gallon (gal)
   2. Determine the Conversion Factor: Converting quarts to gallon involves dividing by 4.
   3. Set Up the Conversion: 48 quarts ÷ 4 = 12 gallons.

Metric Weight and Mass Measurement

Metric weight and mass measurement refers to the system used to measure the weight and mass of objects in the metric system. It uses standardized units such as grams, kilograms, and milligrams.

Conversion Steps:

1. **Understand the Metric Prefixes:**
    f. 1 milligram (mg) = 0.001 grams (g)
    g. 1 gram (g) = 1,000 milligrams (mg)
    h. 1 gram (g) = 0.001 kilograms (kg)
    i. 1 kilogram (kg) = 1,000 grams (g)
    j. 1 kilogram (kg) = 0.001 metric tons (t)
    k. metric ton (t) = 1,000 kilograms (kg)
2. **Determine the Conversion Factor:** Identify the units you are converting from and to.
3. **Set Up the Conversion:** Use the conversion factor to multiply or divide the given value.

Examples:

1) Convert 7,000 milligrams to grams:

Solution:

1. Determine the Starting Unit and Desired Unit:
    - Starting unit: milligram (mg)
    - Desired unit: gram (g)
2. Determine the Conversion Factor: Converting milligrams to grams involves dividing by 1,000.
3. Set Up the Conversion: 7,000 milligrams ÷ 1,000 = 7 grams.

2) Convert 0.003 tons to kilograms:

Solution:

1. Determine the Starting Unit and Desired Unit:
    - Starting unit: tons (t)
    - Desired unit: kilogram (kg)
2. Determine the Conversion Factor: Converting tons to kilograms involves multiplying by 1,000.
3. Set Up the Conversion: 0.003 tons × 1,000 = 3 kilograms.

Customary Weight and Mass Measurement

Customary weight and mass measurement is a system used primarily in the United States to measure the weight and mass of objects. This system includes units such as ounces, pounds, and tons.

Common Conversion Factors:

a. 1 pound (lb) = 16 ounces (oz)

b. 1 ton = 2,000 pounds (lb)

Conversion Process:

1. **Determine the Starting Unit and the Desired Unit**: Identify the units you are converting from and to.

2. **Find the Conversion Factor**: Use the appropriate conversion factor to switch between units.

3. **Multiply or Divide**: Multiply or divide the starting value by the conversion factor.

Examples:

1) Convert 5 pounds to ounces:

Solution:
   1. Determine the Starting Unit and Desired Unit:
      - Starting unit: pound (lb)
      - Desired unit: ounce (oz)
   2. Determine the Conversion Factor: Converting pounds to ounces involves multiplying by 16.
   3. Set Up the Conversion: 5 pounds × 16 = 80 ounces.

2) Convert 3 tons to pounds:

Solution:
   1. Determine the Starting Unit and Desired Unit:
      - Starting unit: tons (t)
      - Desired unit: pound (lb)
   2. Determine the Conversion Factor: Converting tons to pounds involves multiplying by 2,000.
   3. Set Up the Conversion: 3 tons × 2,000 = 6,000 pounds.

Temperature

Temperature is measured using different units depending on the region and context. Here are the main units:

Main Units in Temperature:

1. Celsius (°C): The Celsius scale is based on the freezing and boiling points of water. Water freezes at 0°C and boils at 100°C.

2. Fahrenheit (°F): The Fahrenheit scale sets the freezing point of water at 32°F and the boiling point at 212°F.

Converting from Celsius (°C) to Fahrenheit (°F):

$$°F = \left(°C \times \frac{9}{5}\right) + 32$$

Converting from Fahrenheit (°F) to Celsius (°C):

$$°C = (°F - 32) \times \frac{5}{9}$$

Examples:

1) Convert 25° C to Fahrenheit:

   Solution:

   According to the formula mentioned above, we have:

   $°F = \left(°C \times \frac{9}{5}\right) + 32 = \left(25 \times \frac{9}{5}\right) + 32 = 45 + 32 = 77°$ Fahrenheit.

2) Convert 113° F to Celsius:

   Solution:

   According to the formula mentioned above, we have:

   $°C = (°F - 32) \times \frac{5}{9} = (113 - 32) \times \frac{5}{9} = 81 \times \frac{5}{9} = 45°$ Celsius.

3) Convert -5° C to Fahrenheit:

   Solution:

   According to the formula mentioned above, we have:

   $°F = \left(°C \times \frac{9}{5}\right) + 32 = \left(-5 \times \frac{9}{5}\right) + 32 = -9 + 32 = 23°$ Fahrenheit

www.mathnotion.com

Time

Time is measured using various units, depending on the context and the scale of measurement. Here are the main units of time:

Conversion Steps:

1. **Understand the Common Time Units:**
    a. 1 second (s) = $\frac{1}{60}$ minutes (min)
    b. 1 minutes (min) = 60 seconds (s)
    c. 1 hour (h) = 60 minutes (min)
    d. 1 day = 24 hours (h)
    e. 1 week = 7 days
    f. 1 month ≈ 30 or 31 days (28 or 29 for February)
    g. 1 year = 12 months
    h. 1 year = 48 weeks
    i. 1 year = 365 days
2. **Determine the Conversion Factor:** Identify the units you are converting from and to.
3. **Set Up the Conversion:** Use the conversion factor to multiply or divide the given value.

Examples:

1) Convert 3 hours to minutes:
   Solution:
   1. Determine the Starting Unit and Desired Unit:
      - Starting unit: hour (h)
      - Desired unit: minute (min)
   2. Determine the Conversion Factor: Converting hours to minutes involves multiplying by 60.
   3. Set Up the Conversion: 3 hours × 60 = 180 minutes.

2) Convert 91 days to weeks:
   Solution:
   1. Determine the Starting Unit and Desired Unit:
      - Starting unit: day
      - Desired unit: week
   2. Determine the Conversion Factor: Converting days to weeks involves dividing by 7.

Set Up the Conversion: 91 days ÷ 7 = 13 weeks.

Word Problems

Steps to Solve Measurement Word Problems:

1. **Understand the Problem:** Carefully read the problem to identify what is being asked and highlight or note key details, quantities, and units mentioned.

2. **Identify What You Know:** Write down the given information, including units of measurement and identify the unknown quantity that needs to be calculated.

3. **Choose the Appropriate Formula or Conversion:** Decide which mathematical operations or conversions are needed to solve the problem.

4. **Set Up the Equation:** Write out the equation or series of steps to solve the problem, using the given information.

5. **Perform the Calculations and Check Your Work**: Carefully perform the mathematical operations, ensuring accuracy in each step and finally make sure the answer is reasonable and make sense in the context of the problem.

Example:

Emma makes lemonade and needs 2 liters of lemon juice. She has a measuring cup that holds 250 milliliters. How many times does she need to fill the measuring cup to get 2 liters of lemon juice?

Solution:

1. **Understand the Problem:** Emma needs 2 liters of lemon juice, and the measuring cup holds 250 milliliters.
2. **Identify What You Know:** 1 liter= 1,000 milliliters and 2 liters =2,000 milliliters.
3. **Determine What You Need to Find**: Number of times to fill the measuring cup.
4. **Choose the Appropriate Conversion**: Convert liters to milliliters if necessary.
5. **Set Up the Equation**:
6. Number of times to fill the cup = Total milliliters needed ÷ Capacity of measuring cup:
7. Times = 2,000 mL ÷ 250 mL
8. **Perform the Calculations and Check Your Work:** Times = 2,000 ÷ 250 = 8. So, Emma needs to fill the measuring cup 8 times to get 2 liters of lemon juice.

Worksheets

Metric Length Measurement

✏ Convert to the units:

1) 2.3 km = ☐ m
2) 8,500 mm = ☐ cm
3) 90,000 cm = ☐ m
4) 0.006 m = ☐ mm
5) 860 m = ☐ cm
6) 0.17 mm = ☐ cm
7) 7,500 cm = ☐ m
8) 400,000 cm = ☐ km
9) 54 mm = ☐ m
10) 1.06 m = ☐ km

Customary Length Measurement

✏ Convert to the units:

1) 15 ft = ☐ in
2) 4.8 yd = ☐ ft
3) 1,746 in = ☐ yd
4) 3.24 in = ☐ yd
5) 1.3 mi = ☐ ft
6) 30,000 yd = ☐ mi
7) 0.612 in = ☐ ft
8) 87,120 ft = ☐ mi
9) 15×10^7 yd = ☐ mi
10) 6,000 ft = ☐ in

Metric Capacity Measurement

✏ Convert to the units:

1) 0.504 l = ☐ ml
2) 250 ml = ☐ l
3) 0.0002 kl = ☐ l
4) 6,000 CC = ☐ ml
5) 89 cl = ☐ l
6) 1.2 l = ☐ CC
7) 5.68 dl = ☐ cl
8) 50 kl = ☐ ml
9) 4.213 ml = ☐ CC
10) 45 ml = ☐ cl

Customary Capacity Measurement

✏ Convert to the units:

1) 0.7 gal = ☐ qt
2) 522.4 pt = ☐ qt
3) 672 c = ☐ gal
4) 72.96 fl oz = ☐ c
5) 1.55 pt = ☐ c
6) 12.4 qt = ☐ gal
7) 0.036 qt = ☐ gal
8) 18.2 c = ☐ fl oz
9) 5.8 gal = ☐ pt
10) 2×10^{-5} qt = ☐ pt

www.mathnotion.com

Metric Weight and Mass Measurement

✎ Convert to the units:

1) 0.712 kg = ☐ g
2) 15,000 g = ☐ kg
3) 800 mg = ☐ g
4) 1.2 T = ☐ kg
5) 0.6 g = ☐ mg
6) 0.007 kg = ☐ mg
7) 90 kg = ☐ T
8) 6,500 mg = ☐ kg
9) 5×10^6 mg = ☐ T
10) 2×10^{-8} T = ☐ g

Customary Weight and Mass Measurement

✎ Convert to the units:

1) 12,600 lb = ☐ T
2) 5.2 T = ☐ lb
3) 0.18 lb = ☐ oz
4) 0.006 T = ☐ oz
5) 40,000 oz = ☐ lb
6) 209.92 oz = ☐ lb
7) 0.021 lb = ☐ oz
8) 6.8×10^{-4} T = ☐ lb
9) 2.4×10^6 oz = ☐ T
10) 1.6×10^3 oz = ☐ lb

Temperature

✎ Convert to the units:

1) $35.6\,^0F$ = ☐ ^0C
2) $54.5\,^0F$ = ☐ ^0C
3) $-25.6\,^0F$ = ☐ ^0C
4) $120.2\,^0F$ = ☐ ^0C
5) $60.44\,^0F$ = ☐ ^0C
6) $18.2\,^0C$ = ☐ ^0F
7) $6\,^0C$ = ☐ ^0F
8) $61\,^0C$ = ☐ ^0F
9) $-24\,^0C$ = ☐ ^0F
10) $0\,^0C$ = ☐ ^0F

Time

✎ Convert to the units:

1) 18.5 hr = ☐ min
2) 26 years = ☐ week
3) 0.2 hr = ☐ sec
4) 264 hr = ☐ day
5) 22 weeks = ☐ day
6) 32.25 year = ☐ month
7) 6,480 sec = ☐ min
8) 1.2×10^5 min = ☐ hr
9) 22 weeks = ☐ day
10) 3 years = ☐ hr

Word Problems

✎ Find the answer to each word problem:

1) Maria is organizing a garden and wants to plant a row of flowers along the border. She has 12 flowerpots, and each pot needs to be spaced 30 centimeters apart. If the first and last pots are placed right at the ends of the row, how long will the entire row be in meters?

2) John is decorating his living room and wants to put up a curtain rod. The rod needs to be 6 feet long to cover the window. However, the hardware store only sells curtain rods in inches. How many inches long does the curtain rod need to be?

3) Samantha is planning her day and wants to allocate time for different activities. She spends 1 hour and 45 minutes working on a project, 30 minutes exercising, and 2 hours and 30 minutes reading. If she wants to know the total time spent on these activities in minutes, how many minutes did she spend in total?

4) A chemist is preparing a solution in a laboratory. She needs to mix 750 milliliters of solution A with 1.25 liters of solution B. After mixing, she pours the combined solution into containers that each hold 0.5 liters. How many containers will she need to hold the entire mixture?

5) Tom is filling his backyard pool for a summer party. The pool has a total capacity of 500 gallons. However, his garden hose fills the pool at a rate of 10 gallons per minute. If Tom starts filling the pool at 8:00 AM, at what time will the pool be filled?

6) Katie is traveling from London, where the temperature is measured in Celsius, to New York, where the temperature is measured in Fahrenheit. She checks the weather forecast for both cities and sees that the temperature in London is 15^0C and the temperature in New York is 59^0F She wants to know which city is warmer and by how many degrees, converting the temperatures to the same unit.

7) David is traveling to two different cities in one day. He starts his journey in Sydney, where the temperature is 25^0C, and then flies to New York, where the temperature is 77^0F his flight departs Sydney at 9:00 AM local time and the flight duration is 20 hours. When he arrives in New York, he needs to know the local time as well as how much warmer New York is compared to Sydney, in degrees Celsius.

8) Rachel is a baker, and she needs to prepare a large batch of dough for a party. The recipe requires 5 pounds of flour and 1 gallon of water for each batch. However, her measuring tools are marked in ounces and quarts. If she needs to prepare 3 batches of dough, how many ounces of flour and quarts of water does she need in total?

9) A moving company needs to transport a large shipment of boxes. Each box weighs 35 pounds, and the company has a total of 1,240 boxes for transport. They can use two types of trucks:
   - Type A Truck: Can carry up to 8,000 pounds.
   - Type B Truck: Can carry up to 12,000 pounds.

   The company needs to decide how many Type A and Type B trucks to use, ensuring that the total weight of the boxes does not exceed the capacity of the trucks. If the company decides to use 4 Type A trucks, how many Type B trucks will they need to transport all the boxes?

Answer of Worksheets

Metric Length Measurement

1) 2,300
2) 850
3) 900
4) 6
5) 86,000
6) 0.017
7) 75
8) 4
9) 0.054
10) 1,060

Customary Length Measurement

1) 180
2) 14.4
3) 48.5
4) 0.09
5) 6,864
6) ≈ 17.05
7) 0.051
8) 16.5
9) ≈ 8,522.7
10) 72,000

Metric Capacity Measurement

1) 504
2) 0.25
3) 0.2
4) 6,000
5) 0.89
6) 1,200
7) 56.8
8) 5×10^7
9) 4.213
10) 4.5

Customary Capacity Measurement

1) 2.8
2) 261.2
3) 42
4) 9.12
5) 3.1
6) 3.1
7) 0.009
8) 145.6
9) 46.4
10) 4×10^{-5}

Metric Weight and Mass Measurement

1) 712
2) 15
3) 0.8
4) 1,200
5) 600
6) 7000
7) 0.09
8) 0.0065
9) 0.005
10) 0.02

Customary Weight and Mass Measurement

1) 6.3
2) 10,400
3) 2.88
4) 192
5) 2,500
6) 13.12
7) 0.336
8) 1.36
9) 75
10) 100

Temperature

1) $2°$
2) $12.5°$
3) $\approx -32°$
4) $\approx 49°$
5) $\approx 15.8°$
6) $\approx 64.76°$
7) $\approx 42.8°$
8) $\approx 141.8°$
9) $-11.2°$
10) $32°$

Time

1) 1,110
2) 1,352
3) 720
4) 11
5) 154
6) 387
7) 108
8) 2,000
9) 154
10) 26,280

Word Problems

1) 3.3 meters
2) 72 inches
3) 285 minutes
4) 4 containers
5) 8:50 Am
6) Same temperatures: $59°F$
7) Arrival Time: 2:00 Pm and Temperature Difference: same temperature: $0°C$
8) Flour: 240 ounces and Water: 12 quarts
9) 4 Type A trucks

Chapter 9: Algebraic Expressions

Topics that you will learn in this chapter:

- Find a Rule
- Writing Variable Expressions
- Identify Terms and Coefficients and Simplifying Variable Expressions
- Equivalent Expressions - Properties
  - Distributive Property
  - Commutative Property
  - Associative Property
- Equivalent Expressions - Stirp Model
- Factor Numerical Expressions
- Factor Variable Expressions
- Factor Variable Expressions - Area Model
- Word Problems
- Worksheets

Find a Rule

"Finding a rule" in algebraic expressions is like discovering a pattern or a consistent way to relate numbers. It is a way to describe how to get from one value to another using algebra.

Steps to Find a Rule in Algebraic Expressions:

1. **Identify Variables:** Find the variables with which you are working. Commonly, these are denoted as x(input) and y (output).
2. **Observe Patterns:** Look at the given values or sequences. Notice how each Y value is related to x.
3. **Determine Operations:** Check if the pattern involves addition, subtraction, multiplication, division, or a combination of these.
4. **Evaluate the Rule:** Apply your suspected rule to different values to see if it consistently works.

Steps to Find the Output According to the Given Rule:

1. **Identify your input value (x).**
2. **Plug that input value into the rule.**
3. **Simplify to find the output value (y).**

Examples:

1) Find the rule between x and y:

X	1	2	3	4
y	3	5	7	9

Solution:
1. Identify Variables: x and y.
2. Observe Patterns:
   - When $x = 1, y = 3$
   - When $x = 2, y = 5$
   - When $x = 3, y = 7$
3. Determine Operations: Notice that as x increases by 1, y increases by 2.
4. Create and Test the Rule: The rule might be: $y = 2x + 1$
   - Evaluate it: if $x = 1, y = 2(1) + 1 = 3$ and if $x = 2, y = 2(2) + 1 = 5$.

2) If $y = 3x - 2$ and $x = 5$, find the output(y).

Solution:

1. Identify your input value (x): $x = 5$
2. Plug that input value into the rule: $y = 3 \times 5 - 2$
3. Simplify to find the output value (y): $y = 13$.

Writing Variable Expressions

Crafting variable expressions is like writing a mini math story! Here is how you can go about it:

Steps to Write Variable Expressions:

1. **Identify the Variables:**
   - Variables are usually represented by letters (like x, y or z).
   - They stand for unknown values or quantities that can change.
2. **Determine the Operations:**
   - Decide what mathematical operations (like addition, subtraction, multiplication, or division) you will use.
   - Operations indicate the relationship between the variables and constants (numbers).
3. **Construct the Expression**: Put the variables, constants, and operations together to form the expression.

Example:

Write a variable expression for the following phrases:

a) The sum of six and a number.
b) Subtract the product of a and b from forty-one.

Solution:

Part a:

1. **Identify the variable:** The phrase "a number" can be represented by a variable, typically x or any other letter you choose.
2. **Determine the Operation**: "The sum of" indicates addition.
3. **Construct the Expression**: Combine the constant (6) with the variable: $6 + x$.

Part b:

1. **Identify the Variables**: a and b are the variables.
2. **Determine the Operations**:
   - "Product of a and b" means $a \times b$.
   - "Subtract the product of a and b from 41" means you need to take 41 and subtract $a \times b$.
3. **Construct the Expression**: $41 - (a \times b) = 41 - ab$.

Identify Terms and Coefficients

Terms: In an algebraic expression, terms are the separate parts of the expression that are added or subtracted.

Coefficients: The coefficient is the numerical part of a term that multiplies the variable.

Example:

✹ Write terms and coefficients in the following expression: $2x + 5y^2 - 7y + 2$

Solution:

- Writing Terms: In this expression we have 4 different terms: $2x$, $5y^2$, $-7y$ and 2.
- Writing Coefficients:
    - For the term $2x$, the coefficient is 2.
    - For the term $5y^2$, the coefficient is 5.
    - For the term $-7y$, the coefficient is -7.
    - Term 2 is a constant term and has no variable.

Simplifying Variable Expressions

Steps to Simplify Variable Expressions:

1. **Combine Like Terms:** Like terms are terms that have the same variable raised to the same power.
2. **Write the Simplified Expression:** Add or subtract the coefficients of the like terms.

Example:

✹ Simplify $4x + 3y - 2x + 7 - y + 2$:

Solution:

1. Combine Like Terms:
    - Combine the x terms: $4x - 2x = 2x$
    - Combine the y terms: $3y - y = 2y$
    - Combine the constant terms: $7 + 2 = 9$
2. Write the Simplified Expression:
    $2x + 2y + 9$

Equivalent Expressions - Properties

Equivalent expressions are different expressions that represent the same value. They might look different but when simplified, they have the same result. For evaluating expressions, we should use different properties:

1. **Distributive Property**: The distributive property is like spreading out multiplication over addition or subtraction within parentheses. It helps you simplify expressions. The property states that for any numbers a, b, and c: $a(b + c) = ab + ac$.

2. **Commutative Property:**
   - **Commutative Property of Addition:** This property states that changing the order of the numbers being added does not change the sum: $a + b = b + a$.
   - **Commutative Property of Multiplication**: This property states that changing the order of the numbers being multiplied does not change the product: $a \times b = b \times a$.

3. **Associative Property:**
   - **Associative Property of Addition:** This property tells us that the way numbers are grouped when adding doesn't affect the sum: $(a + b) + c = a + (b + c)$.
   - **Associative Property of Multiplication**: This property tells us that the way numbers are grouped when multiplying doesn't affect the product: $(a \times b) \times c = a \times (b \times c)$.

Steps to Write Equivalent Expressions:

1. **Use appropriate property.**
2. **Combine like terms.**

Example:

Use appropriate property to simplify $(3x - 2)(-8) + 4(x + 5)$.

Solution:

1. Using distributive and commutative properties:

   $(3x - 2)(-8) = (-8)(3x - 2) = (-8) \times 3x + (-8) \times (-2) = -24x + 16$
   $4(x + 5) = 4 \times x + 4 \times 5 = 4x + 20$

2. Combining like terms:

   $-24x + 16 + 4x + 20 = -24x + 4x + 16 + 20 = -20x + 36$.

Evaluating Variable Expressions

Evaluating variable expressions is all about *substituting* the right values and carefully *simplifying*.

Steps to Evaluate Variable Expressions:

1. **Identify the Variables:** Recognize the variables in the expression, often represented by letters like x, y, etc.
2. **Substitute the Values:** Replace the variables with the given values.
3. **Simplify the Expression**: Perform the arithmetic operations to simplify the expression (Follow Order of Operations).
4. **Check Your Work**: Simple mistakes often come from rushing. Double-check each step.

Examples:

1) Evaluate $9x^2 + 3x - 6, x = 2$ using the value given:

   Solution:
   1. Substitute $x = 2$ into the expression:
      $9(2^2) + 3(2) - 6$
   2. Perform the multiplication and simplify the expression:
      $9 \times (2^2) + 3 \times (2) - 6 = 9 \times 4 + 6 - 6 = 36 + 6 - 6 = 36$

2) Evaluate $\frac{5x-9}{3y^2-1}, x = 3 \text{ and } y = -1$ using the values given:

   Solution:
   1. Substitute $x = 3 \text{ and } y = -1$ into the expression:
      $\frac{5(3)-9}{3(-1)^2-1}$
   2. Calculate the numerator:
      $5 \times 3 - 9 = 15 - 9 = 6$
   3. Calculate the denominator:
      $3 \times (-1)^2 - 1 = 3 \times 1 - 1 = 3 - 1 = 2$
   4. Simplify the expression:
      $\frac{5(3)-9}{3(-1)^2-1} = \frac{6}{2} = 3$

Equivalent Expressions - Strip Model

A strip model (or bar model) is a rectangular diagram that you can split into sections to represent different parts of a math problem. In algebra, we often use strip models to represent an expression like $x + 3$ or $2x + 5$ by cutting a bar into x-sections (unknown parts) and fixed (constant) sections.

Also, strip model is a visual representation used to show relationships between different parts of an expression. It's especially helpful in understanding and finding equivalent expressions. Let's break it down:

Steps to Use a Strip Model:

1. **Identify the Expression**: Start with the expression you want to visualize.
2. **Create Strips for Each Term:** Draw a strip for each part of the expression.
3. **Label and Combine:**
   - Label each part of the strip with the corresponding term.
   - Combine the strips to show how the terms add up to form the expression.

Example:

Simplify $-2(2x + 3) + 4(x - 5)$ using strip model:

Solution:

First Part: $-2(2x + 3)$

1. Draw the Strip: Imagine a strip divided into $2x$ and 3. This strip is then repeated twice (since it's multiplied by 2), but with a negative sign.
2. Visualize and Combine: You have $-2 \times 2x = -4x$ and $-2 \times 3 = -6$.

Second Part: $4(x - 5)$

1. Draw the Strip: Imagine a strip divided into x and -5. This strip is then repeated four times (since it's multiplied by 4).
2. Visualize and Combine: You have $4 \times x = 4x$ and $4 \times -5 = -20$.

Combine Both Parts: $-4x - 6 + 4x - 20$

Simplify: $-4x - 6 + 4x - 20 = -26$

Factor Numerical Expressions

Factoring numerical expressions using the distributive property is like reversing distribution. Here are the suggested steps to take:

Steps:

1. **Identify the Common Factor**: Look for a number that can divide each term in the expression.
2. **Rewrite the Expression**: Factor out the common number from each term.
3. **Use the Distributive Property**: Write the expression as a product of the common factor and the simplified expression within parentheses.

Example

Factor below expression using the distributive property:

$48 + 64 + 80$

Solution:

1. Identify the Common Factor: The greatest common factor (GCF) for 48, 64, and 80 is 16.
2. Rewrite the Expression:
   - $48 = 16 \times 3$
   - $64 = 16 \times 4$
   - $80 = 16 \times 5$
3. Use the Distributive Property:
   $48 + 64 + 80 = 16(3) + 16(4) + 16(5)$
4. Use the Distributive Property:
   $16(3 + 4 + 5)$
5. Simplify inside the parentheses:
   $3 + 4 + 5 = 12$
6. Final Expression:
   $48 + 64 + 80 = 16 \times 12 = 16(12)$

www.mathnotion.com

Factor Variable Expressions

Let's dive into factoring variable expressions using the distributive property. It's like reversing the distributive process. Here's a step-by-step guide:

Steps:

1. **Combine Like Terms**
2. **Identify the Common Factor:**
   - Look for the highest common factor (HCF) or the greatest common factor (GCF) in each term.
   - This factor could be a number, a variable, or a combination of both.
3. **Rewrite Each Term:** Express each term as a product of the common factor and another term.
4. **Factor Out the Common Factor**: Use the distributive property in reverse to write the expression as the product of the common factor and the simplified expression within parentheses.

Example:

Factor $4x^3 - 12x + 20x^2 + 12x^3$ using the distributive property:

Solution:

1. **Combine Like Terms**:
    - Combine the x^3 terms: $4x^3 + 12x^3 = 16x^3$.
    - The expression now is: $16x^3 + 20x^2 - 12x$

2. **Identify the Common Factor**:
    - The common factor for $16x^3$, $20x^2$, and $-12x$ is $4x$.

3. **Rewrite Each Term**:
    - $16x^3 = 4x \times 4x^2$
    - $20x^2 = 4x \times 5x$
    - $-12x = 4x \times (-3)$

4. **Factor Out the Common Factor**:

   $16x^3 + 20x^2 - 12x = 4x(4x^2 + 5x - 3)$

Factor Variable Expressions - Area Model

The area model is a visual method to factor variable expressions, especially useful when dealing with polynomials. It breaks down the expression into smaller, more manageable pieces.

Steps to Use the Area Model:

1. **Draw a Rectangle:** Divide the rectangle into sections based on the terms of the expression.
2. **Label the Sections:** Each section represents a term from the expression.
3. **Fill in the Factors:** Write the factors on the sides of the rectangle.
4. **Multiply to Find the Areas:** Multiply the labels to find the product in each section.
5. **Combine the Sections:** Add the areas of all sections to form the factored expression.

Quick Tips for the Area Model

1. Count the Terms:
    - 2 terms: Likely you're factoring out a *GCF* (like $2x + 6$).
    - 3 terms: Possibly a simple *trinomial* (like $x^2 + 5x + 6$).
2. GCF First: Always check if there's a GCF you can factor out before you do anything else.
3. Use the Box to Organize: Each *term* of the expression goes into its own box. Then factor out the greatest common factor from rows and columns.
4. Check: Multiply your factored expression to make sure it matches the original.

Example:

★ Factor $8x + 24x^2 - 36x^3$ using area model.

Solution:

1. Draw a Rectangle and Label the Sections: Each section represents a term from the expression.

$8x$	$24x^2$	$-36x^3$

2. Fill in the Factors: The common factor for $8x$, $24x^2$ and $-36x^3$ is $4x$ and the other sides of each section are 2, 6x and $-9x^2$:

	2	6x	$-9x^2$
$4x$	$8x$	$24x^2$	$-36x^3$

3. Multiply to Find the Areas:
    - $4x \times 2 = 8x$
    - $4x \times 6x = 24x^2$
    - $4x \times (-9)x^2 = -36x^3$

4. Combine the Sections: Add the areas of all sections to form the factored expression:
$8x + 24x^2 - 36x^3 = 4x(2 + 6x - 9x^2)$

Word problems

Solving word problems with algebraic expressions can be a fun way to apply math to real-life situations.

Steps to Solve Word Problems:

1. **Read the Problem Carefully**: Understand what the problem is asking. Identify the quantities involved and what needs to be found.
2. **Identify the Variables:** Assign variables to unknown quantities. This could be something like x, y, etc.
3. **Translate Words into an Algebraic Expression**: Convert the word problem into an algebraic expression or equation using the variables.
4. **Set Up the Equation**: Based on the information given in the problem, set up an equation.
5. **Solve the Equation**: Use algebraic methods to solve the unknown variables.
6. **Check Your Solution:** Plug the solution back into the original word problem to see if it makes sense.

Example:

Sunny earns $12 per hour delivering cakes. She worked for x hours this week. Unfortunately, she was charged $15 for a late delivery on Tuesday. How much money did Sunny earn this week? Write your answer as an expression.

Solution:

Identify the variable: Sunny earns $12 per hour, and she worked x hours this week.

Translate words into an algebraic expression:

- Gross Earnings = $12x$
- Sunny was charged $15 for a late delivery, so we need to subtract this amount from her gross earnings.

Set up the equation: Net earning = gross earning – late delivery charge: $12x - 15$

Worksheets

Find a Rule

✍ Find the rule between x and y:

1)

x	1	2	3	4
y	5	10	15	20

2)

x	1	2	3	4
y	3	5	7	9

3)

x	1	2	3	4
y	1	4	9	16

4)

x	1	2	3	4
y	2	7	12	17

5)

x	1	2	3	4
y	0	3	8	15

6)

x	1	2	3	4
y	$\frac{5}{7}$	$\frac{5}{11}$	$\frac{5}{15}$	$\frac{5}{19}$

7)

x	1	2	3	4
y	2	6	12	20

8)

x	1	2	3	4
y	1	$\frac{1}{4}$	$\frac{1}{9}$	$\frac{1}{16}$

9)

x	1	2	3	4
y	1	-1	-3	-5

10)

x	1	2	3	4
y	$\frac{2}{3}$	$\frac{4}{8}$	$\frac{6}{13}$	$\frac{8}{18}$

Writing Variable Expressions

✍ Write a variable expression for the following phrases:

1) 9 multiplied by x
2) Subtract 11 from y
3) 38 decreased by y
4) The difference between fifty–seven and y.
5) The difference between x and 6 is 19
6) The square of z
7) Two units less than the square of y
8) The sum of double and the square of a.
9) The difference between half and 3 times of x
10) The difference between the square of b and 5

Identify Terms and Coefficients

✍ Write terms and coefficients in the following expression:

1) $5x - 1$
2) $2y^2 + 9y + 5$
3) $4x^2 - 8x - 2x$
6) $-4 - 2x^2 - 6x^2$
7) $0.2x + 8x^4$
8) $4x - 5 + 15x + 3x^2$

4) $9a^4 - 0$

5) $\frac{8}{10}b$

9) $\frac{3y}{4} - \frac{5}{6}y^2 - \frac{1}{3}$

10) $\frac{8-x}{2}$

Simplifying Variable Expressions

✍ simplify following expression:

1) $3x - 5x$
2) $4y^2 + 2y - y^2 + 8y$
3) $8x^2 - 4x + 8 - 2x - 9$
4) $24 - 3a + 6b - a^2 + 2b - 3 + 9a$
5) $3xy - y + 2x - 7xy + 5y + x$

Write Equivalent Expressions Using Properties

✍ simplify each expression using properties:

1) $3(x + 5)$
2) $3x^2 + 2(x + 15x^2)$
3) $6x + 7(3 - 4x)$
4) $6(-3x - 9) - 17$
5) $(-11)(-5x + 2) - 41x$

6) $16x - 3x(x + 10)$
7) $10x - 9x^2 - 3x^2 - 7(x - 2)$
8) $-2(3y - 5y^2) + 4(y^2 + 5y)$
9) $17x(1 + x) + 5x(2 - 4x)$
10) $2a(-a^3 + a - 8) - (-a^2)$

Equivalent Expressions Using Stirp Model

✍ simplify each expression using strip model:

1) $2(x + 3)$
2) $4(b + 2) + b$
3) $6(x + 2) + 3$
4) $7(y + 3) + y$
5) $3(4x + 1) + 2x$

6) $2(5y - 2) + 3y$
7) $2(x + 5) + 3(x - 2)$
8) $4(a + 2) + 5(a - 3)$
9) $-2(z - 6) - 5(z + 1)$
10) $4(a + 2 - b) - 2(b - 2a + 1)$

Factor Numerical Expressions

✍ Factor below expression using the distributive property:

1) $6 + 3$
2) $12 + 9$
3) $24 + 16$
4) $10 + 20 + 30$
5) $30 + 45$

6) $16 + 12 + 18$
7) $48 + 36 + 72$
8) $30 + 27 + 81$
9) $150 + 120 + 90$
10) $75 + 60 + 45 + 150$

Factor Variable Expressions

✍ Factor below expression using the distributive property:

1) $2x + 4$
2) $7a + 14a^2$
3) $10d^2 + 15d - 5d^2$
4) $12x - 6x^2 + 8$
5) $20a^3 - 15a^2 + 10a^4$
6) $25b - 10b^2 + 5b + 35b^2$
7) $4x^2 + 25x + 21x^2$
8) $(-8)(3x - 2) + 4(x + 5)$
9) $2y(x - 2) + 4(x - 2)$
10) $(-6)(10x - 4) - 2(5x - 2)$

Factor Variable Expressions - Area Model

✍ Factor following expressions using area model.

1) $5x - 10$
2) $2x - 4x^2$
3) $6y^2 - 15y$
4) $9z^3 + 21z$
5) $2x^2 + 8x - 4$
6) $15y^2 - 10y + 5$
7) $20a^2 - 30a + 10$
8) $24x^3 + 36x^2 - 12x$
9) $30y^3 - 45y^2 + 15y$
10) $10b - 2b^3 + 6b^2$

Word Problems

✍ write an algebraic equation.

1) Lisa has twice as many pencils as Tom. Together, they have 18 pencils. How many pencils does each person have?
2) A garden has a length that is 5 meters longer than its width. If the perimeter of the garden is 30 meters, find the length and width of the garden.
3) A movie ticket costs $8, and a bucket of popcorn costs $5. If a family spends $51 on movie tickets and popcorn, how many tickets and buckets of popcorn did they buy?
4) The rectangle's area is 60 square meters. If the length is 4 meters longer than the width, find the dimensions of the rectangle.
5) The sum of the ages Alice and Bob is 30 years. Alice is 4 years older than Bob. How old are Alice and Bob?
6) A company sells two types of candies: regular candies for $3 per bag and premium candies for $5 per bag. If they sold a total of 100 bags and made $400, how many bags of each type did they sell?
7) A car rental company charges $20 per day plus a one-time fee of $50. If a customer paid $110, how many days did they rent the car?
8) The number of boys in a class is 3 more than twice the number of girls. If there are 21 students in total, how many boys and girls are there?
9) Emma bought 3 notebooks and 2 pens for $14. She bought another notebook and a pen for $5. Find the cost of one notebook and one pen.

Answer of Worksheets

Find a Rule

1) $y = 5x$
2) $y = 2x + 1$
3) $y = x^2$
4) $y = 5x - 3$
5) $y = x^2 - 1$
6) $y = \frac{5}{4x+3}$
7) $y = x(x + 1)$
8) $y = \frac{1}{x^2}$
9) $y = -2x + 3$
10) $y = \frac{2x}{5x-2}$

Writing Variable Expressions

1) $9x$
2) $y - 11$
3) $y - 38$
4) $57 - y$
5) $x - 6 = 19$
6) z^2
7) $y^2 - 2$
8) $2a + a^2$
9) $\frac{x}{2} - 3x$
10) $b^2 - 5$

Identify Terms and Coefficients

1) Terms: $5x$ and -1
   Coefficient: 5
2) Terms: $2y^2$, $9y$ and 5
   Coefficient: 2, 9
3) Terms: $4x^2, -8x$ and $-2x$
   Coefficients: 4, -8 and -2
4) Term: $9a^4$
   Coefficient: 9
5) Term: $\frac{8}{10}b$
   Coefficient: $\frac{8}{10}$
6) Terms: $-4, -2x^2$ and $-6x^2$
   Coefficients: -4, -2 and -6
7) Terms: $0.2x$ and $8x^4$
   Coefficients: 0.2 and 8
8) Terms: $4x, -5, 15x$ and $3x^2$
   Coefficients: 4, 15 and 3
9) Terms: $\frac{3y}{4}, -\frac{5}{6}y^2$ and $-\frac{1}{3}$
   Coefficients: $\frac{3}{4}$ and $-\frac{5}{6}$
10) Terms: $\frac{8}{2}$ and $\frac{-x}{2}$
    Coefficient: $\frac{-1}{2}$

Simplifying Variable Expressions

1) $-2x$
2) $3y^2 + 10y$
3) $8x^2 - 6x - 1$
4) $21 + 6a + 8b - a^2$
5) $-4xy + 3x + 4y$

Write Equivalent Expressions Using Properties

1) $3x + 15$
2) $33x^2 + 2x$
3) $-22x + 21$
4) $-18x - 71$
5) $14x - 22$
6) $-14x - 3x^2$
7) $3x - 12x^2 + 14$
8) $14y + 14y^2$
9) $27x - 3x^2$
10) $-2a^4 - 16a + 3a^2$

Equivalent Expressions Using Stirp Model

1) $2x + 6$
2) $5b + 8$
3) $6x + 15$
4) $8y + 21$
5) $14x + 3$
6) $13y - 4$
7) $5x + 4$
8) $9a - 7$
9) $-7z + 7$
10) $8a - 6b + 6$

Factor Numerical Expressions

1) 3(3)
2) 3(7)
3) 8(5)
4) 10(6)
5) 15(5)
6) 2(23)
7) 12(13)
8) 3(46)
9) 30(12)
10) 15(22)

Factor Variable Expressions

1) $2(x + 2)$
2) $7a(1 + 2a)$
3) $5d(d + 3)$
4) $2(6x - 3x^2 + 4)$
5) $5a^2(4a - 3 + 2a^2)$
6) $5b(6 + 5b)$
7) $25x(x + 1)$
8) $4(-5x + 9)$
9) $(x - 2)(2y + 4)$
10) $-14(5x - 2)$

Factor Variable Expressions: Area Model

1)
	x	-2
5	$5x$	-10

2)
	1	$-2x$
$2x$	$2x$	$-4x^2$

3)
	$2y$	-5
$3y$	$6y^2$	$-15y$

4)
	$3z^2$	7
$3z$	$9z^3$	$21z$

5)
	$2x$	$4x$	-2
2	$2x^2$	$8x$	-4

6)
	$3y^2$	$-2y$	1
5	$15y^2$	$-10y$	5

7)
	$2a^2$	$-3a$	1
10	$20a^2$	$-30a$	10

8)
	$2x^2$	$3x$	-1
$12x$	$24x^3$	$36x^2$	$-12x$

9)
	$2y^2$	$-3y$	1
$15y$	$30y^3$	$-45y^2$	$15y$

10)
	5	$-b^2$	$3b$
$2b$	$10b$	$-2b^3$	$6b^2$

Word Problems

1) $x + 2x = 18$
2) $2w + 2(w + 5) = 30$
3) $8x + 5y = 51$
4) $w(w + 4) = 60$
5) $a + b = 30$ and $a = b + 4$
6) $3x + 5y = 400$ and $x + y = 100$
7) $20x + 50 = 110$
8) $b = 2g + 3$ and $b + g = 21$
9) $3x + 2y = 14$ and $x + y = 5$

Chapter 10: Equations and Inequalities

Topics that you will learn in this chapter:

- Identify Expressions and Equations
- One-Step Equations
- Two-Step Equations
- Multi-Step Equations
- Equation Diagram
- One-Step Inequalities
- Two-Step Inequalities
- Multi-Step Inequalities
- Graphing Inequalities
- Word Problems
- Worksheet

Identify Expressions and Equations

Understanding the difference between expressions and equations is a key part of algebra. As we taught in the previous chapter, an **expression** is a combination of numbers, variables, and operators (like +, -, ×, and ÷) that represents a value. It **does not** have an equal sign. However, an **equation** is a mathematical statement that two expressions are equal. It includes an equal sign (=). So, to identify expressions and equations we can note the steps below:

1. **Look for the Equal Sign**:
   - **Equations** have an equal sign (=).
   - **Expressions** do not have an equal sign.

2. **Structure**:
   - **Expressions** are usually shorter and represent a value or a calculation.
   - **Equations** represent a relationship between two expressions.

3. **Purpose**:
   - **Expressions** are used to describe a value or a calculation.
   - **Equations** are used to find the value of unknown variables by solving them.

Example:

Identify whether each is an expression or an equation:

a) $5x + 3 = 2x + 8$

b) $7y - 4$

c) $12 + 6 = 18$

d) $4a^2 - 3a + 2$

Solution:

Equations have an equal sign and expressions do not have an equal sign:

a) Equation

b) Expression

c) Equation

d) Expression

One–Step Equations

A one-step equation is an algebraic equation that can be solved in a single step, either by addition, subtraction, multiplication, or division. Here's a step-by-step guide on how to solve a one-step equation:

Steps to Solve a One-Step Equation:

1. **Identify the Operation:** Determine which operation (addition, subtraction, multiplication, or division) is being used in the equation.
2. **Perform the Inverse Operation**: To isolate the variable, perform the inverse (or opposite) operation on both sides of the equation.
3. **Simplify Both Sides:** Simplify both sides of the equation to find the value of the variable.

Common Mistakes to Avoid:

- Forgetting to **do the same thing to both sides** of the equation.
- Mixing up addition and subtraction or multiplication and division.
- Not checking your answer by plugging it back in.

Examples:

Solve.

a) $x + 5 = 12$

b) $12 = -2z$

Solution:

Part a:

1. Identify the Operation: Addition $(+5)$
2. Perform the Inverse Operation: Subtract 5 from both sides of the equation:
$$x + 5 - 5 = 12 - 5$$
3. Simplify Both Sides: $x = 7$

Part b:

1. Identify the Operation: Multiply (-2)
2. Perform the Inverse Operation:
Divide both sides by (-2): $\frac{12}{-2} = \frac{-2z}{-2}$
3. Simplify Both Sides: $z = -6$

Two-Step Equations

Solving a two-step equation involves isolating the variable by undoing two operations, usually in the order of reverse operations. Here's a step-by-step guide:

Steps to Solve a Two-Step Equation:

1. **Identify the Operations**: Determine the two operations that are affecting the variable.
2. **Perform the Inverse Operations:**
   - Start with the inverse of the addition or subtraction operation.
   - Then use the inverse of the multiplication or division operation.
3. **Isolate the Variable:** Ensure the variable is on one side of the equation and solve it.

Common Mistakes to Avoid

- Doing multiplication or division before addition or subtraction.
- Forgetting to **do the same operation on both sides** of the equation.
- Skipping the checking step.

Examples:

Solve two-steps equations:

a) $3x + 4 = 19$

b) $5(2y - 1) = -15$

Solution: part a:

1. Identify the Operations: The equation has addition (+4) and multiplication (3 times x).
2. Perform the Inverse Operations:
   - Step 1: Subtract 4 from both sides to undo the addition:
     $3x + 4 - 4 = 19 - 4$ so, $3x = 15$
   - Step 2: Divide both sides by 3 to undo the multiplication: $\frac{3x}{3} = \frac{15}{3}$ so, $x = 5$

Part b: First we must distribute 5 to both items inside the parentheses: $5(2y - 1) = 10y - 5$ so we have $10y - 5 = -15$

1. Identify the Operations: The equation has subtraction (−5) and multiplication (10 times y).
2. Perform the Inverse Operations:
   - Step 1: add 5 to both sides to undo the subtraction:
     $10y - 5 + 5 = -15 + 5$ so, $10y = -10$
   - Step 2: Divide both sides by 10 to undo the multiplication: $\frac{10y}{10} = \frac{-10}{10}$ so, $y = -1$

Multi−Step Equations

Solving multi-step equations involves more steps compared to one-step or two-step equations. Here's a detailed guide to help you solve multi-step equations:

Steps to Solve Multi-Step Equations:

1. **Simplify Both Sides:**
    - Combine like terms on each side of the equation.
    - Use the distributive property to remove parentheses, if necessary.
2. **Move Variable Terms to One Side:** Use addition or subtraction to get all variable terms on one side of the equation.
3. **Move Constant Terms to the Other Side**: Use addition or subtraction to get all constant terms on the opposite side of the equation from the variable terms.
4. **Combine Like Terms:** Simplify both sides of the equation again, if needed.
5. **Isolate the Variable:** Use multiplication or division to solve for the variable.
6. **Check Your Solution:** Substitute the solution back into the original equation to ensure it satisfies the equation.

Common Mistakes to Avoid

- ✘ Forgetting to distribute correctly.
- ✘ Not combining like terms first.
- ✘ Moving variables to different sides incorrectly.
- ✘ Skipping the checking step.

Example:

Solve the equation $2(x + 3) = 4x − 6$.

Solution:

1. Simplify both Sides: Distribute the 2:
$$2x + 6 = 4x − 6$$
2. Move Variable Terms to One Side: Subtract 2x from both sides:
$2x + 6 − 2x = 4x − 6 − 2x$ and now we have: $6 = 2x − 6$
3. Move Constant Terms to the Other Side: Add 6 to both sides:
$6 + 6 = 2x − 6 + 6$ and then $12 = 2x$
4. Isolate the Variable: Divide both sides by 2:
$\frac{12}{2} = \frac{2x}{2}$ and finally, $6 = x$.

Equation Diagram

An equation diagram is a visual representation used to illustrate mathematical equations or relationships between different variables. It typically involves geometric shapes, arrows, and labels to show how the elements of an equation are connected. Here are steps to solve equations with a simple kind of diagram:

1. **Identify the Equation that Represents the Diagram.**
2. **Solve the Equation:**
   - Identify the Operations: Determine the two operations that are affecting the variable.
   - Perform the Inverse Operations: Start with the inverse of the addition or subtraction operation. Then, use the inverse of the multiplication or division operation.
   - Isolate the Variable: Ensure the variable is on one side of the equation and solve it.

Example:

Write an equation for each diagram.

a) 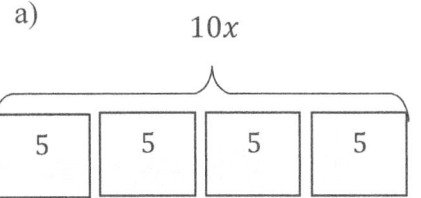   b)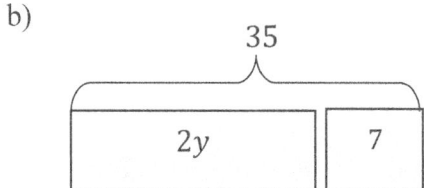

Solution:

a)
1) Identify the Equation that Represents the Diagram: This diagram says If we add the number 5 to itself 4 times or multiply it by 4, the result equals $10x$. So, the equation is: $10x = 5 \times 4$ or $10x = 20$.
2) Solve the Equation: Divide both sides by 10 and then simplify both sides:
$$\frac{10x}{10} = \frac{20}{10}$$
$$x = 2.$$

b)
1) Identify the Equation that Represents the Diagram: This diagram says if we add $2y$ to 7, the result equal to 35. So, the equation is: $2y + 7 = 35$.
2) Solve the Equation: Subtract 7 from both sides and then divide them by 2:
$$2y + 7 - 7 = 35 - 7$$
$$2y = 28$$
$$\frac{2y}{2} = \frac{28}{2} \rightarrow y = 14.$$

One-Step Inequalities

One-step inequalities are like one-step equations, but instead of an equal sign, they involve inequality signs ($<, \leq, >, \geq$). Solving one-step inequalities follows almost the same process as solving one-step equations. Here's a detailed guide to help you solve them:

Steps to Solve One-Step Inequalities:

1. **Identify the Operation:** Determine the operation being used (addition, subtraction, multiplication, or division).
2. **Perform the Inverse Operation:** Use the inverse operation to isolate the variable on one side of the inequality.
3. **Solve the Inequality:**
    - Simplify finding the solution.
    - Remember that if you multiply or divide by a negative number, you must reverse the inequality sign.

Example:

Solve.

a) $x + 5 < 12$

b) $4y \geq 16$

c) $\frac{a}{-3} > 2$

Solution: part a:

3. Identify the Operation: Addition (+5)
4. Perform the Inverse Operation: Subtract 5 from both sides: $x + 5 - 5 < 12 - 5$ so, $x < 7$

Part b:

1. Identify the Operation: Multiplication ($\times 4$)
2. Perform the Inverse Operation: Divide both sides by 4: $\frac{4y}{4} \geq \frac{16}{4}$ so, $y \geq 4$

Part c:

1. Identify the Operation: Division ($\div -3$)
2. Perform the Inverse Operation: Multiply both sides by -3. Remember to reverse the inequality sign: $a < 2 \times -3$ so, $a < -6$

Two-Steps Inequalities

Two-step inequalities are like one-step inequalities but require two steps to isolate the variable. These inequalities involve more than one operation (addition, subtraction, multiplication, or division). Here's a step-by-step guide to solve them:

Steps to Solve Two-Step Inequalities:

1. **Simplify Both Sides (if necessary):** Combine like terms on each side of the inequality, if needed.
2. **Perform the Inverse Operation for Addition/Subtraction:** Use addition or subtraction to move the constant term to the other side of the inequality.
3. **Perform the Inverse Operation for Multiplication/Division:** Use multiplication or division to isolate the variable. Remember that if you multiply or divide by a negative number, you must reverse the inequality sign.
4. **Graph the Solution:** Represent the solution on a number line.

Example:

Solve.

a) $3x + 5 \leq 14$

b) $2(1 - y) > 6$

Solution:

Part a:

1. Perform the Inverse Operation for Subtraction: Subtract 5 from both sides to move the constant term: $3x + 5 - 5 \leq 14 - 5$ so, $3x \leq 9$
2. Perform the Inverse Operation for Multiplication: Divide both sides by 3 to isolate x: $\frac{3x}{3} \leq \frac{9}{3}$ so, $x \leq 3$.

Part b:

1. Simplify Left Side: Distribute the 2: $2 - 2y > 6$
2. Perform the Inverse Operation for Subtraction: Subtract 2 from both sides to move the constant term: $2 - 2y - 2 > 6 - 2$ so, $-2y > 4$
3. Perform the Inverse Operation for Multiplication: Divide both Sides by -2 to isolate y. Remember to reverse the inequality sign: $\frac{-2y}{-2} < \frac{4}{-2}$ so, $y < -2$.

Multi-Step Inequalities

Multi-step inequalities are inequalities that require more than one operation to isolate the variable. These inequalities might involve a combination of addition, subtraction, multiplication, division, and distribution, like multi-step equations but with inequalities signs ($<, >, \leq, \geq$) instead of an equal sign.

Steps to Solve Multi-Step Inequalities:

1. **Simplify Both Sides:**
   - Combine like terms on each side of the inequality.
   - Use the distributive property to remove parentheses if necessary.
2. **Move Variable Terms to One Side:** Use addition or subtraction to get all variable terms on one side of the inequality.
3. **Move Constant Terms to the Other Side:** Use addition or subtraction to get all constant terms on the opposite side from the variable terms.
4. **Combine Like Terms:** Simplify both sides of the inequality again, if needed.
5. **Isolate the Variable**: Use multiplication or division to isolate the variable. Remember to reverse the inequality sign if you multiply or divide by a negative number.

Example:

Solve the multi-step inequality $3(x - 2) + 4 \leq 2(x + 1)$.

Solution:

1. Simplify Both Sides: Distribute the 3 on the left-hand side and the 2 on the right-hand side:
   $3(x - 2) + 4 \leq 2(x + 1)$
   $3x - 6 + 4 \leq 2x + 2$
   $3x - 2 \leq 2x + 2$
2. Move Variable Terms to One Side: Subtract $2x$ from both sides to move variable terms to one side:
   $3x - 2x - 2 \leq 2x - 2x + 2$
   $x - 2 \leq 2$
3. Move Constant Terms to the Other Side: Add 2 to both sides to move the constant term:
   $x - 2 + 2 \leq 2 + 2$
   $x \leq 4$.

Graphing Inequalities

Graphing inequalities on a number line is a straightforward process that visually represents the range of possible solutions. Here's a step-by-step guide:

Steps to Graph Inequalities on Number Lines:

1. **Understand the Inequality:** Identify whether the inequality is $<, \leq, >,$ or $\geq$.
2. **Plot the Critical Point:**
    - Find the number that the variable is being compared to (the critical point).
    - Decide whether to use an open circle (for $<$ or $>$) or a closed circle (for $\leq$ or $\geq$).
3. **Shade the Appropriate Region:**
    - Shade the region of the number line that represents the solutions to the inequality.
    - For $<$ or $\leq$, shade to the left.
    - For $>$ or $\geq$, shade to the right.

    Check a Test Point: If you're unsure which side to shade, pick a number on one side of the boundary and see if it makes the inequality **true**. If yes, shade that side; if not, shade the other side.

Example:

✱ Solve inequality $x - 5 > -2$ and then graph the Solution:

Solution:

1. According to the steps for solving inequalities we must add 5 to both sides of inequality:
$x - 5 + 5 > -2 + 5$
$x > 3$
2. Identify the Inequality: The inequality is $x > 3$.
3. Plot the Critical Point: Use an open circle at 3 since it is a strict inequality ($>$).

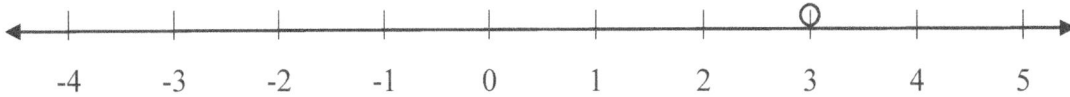

4. Shade the region to the right of 3:

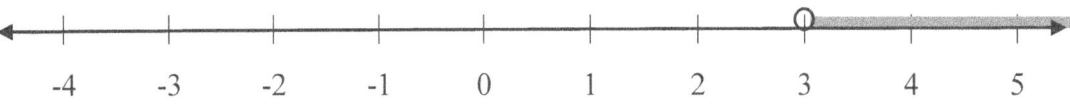

Word Problems

The process of solving word problems about equations and inequalities is almost the same as previous chapters:

Step-by-Step Guide to Solving Word Problems:

1. **Read the Problem Carefully:** Understand what is being asked and identify key information. Sometimes rephrasing the problem in simpler terms can help.
2. **Identify Variables:** Assign variables (like x or y) to the unknown quantities mentioned in the problem.
3. **Write Down What You Know:** Translate the word problem into a mathematical equation or inequality. Make sure every part of the problem is accounted for.
4. **Formulate the Equation or Inequality**: Based on the relationships described in the problem, write down one or more equations or inequalities.
5. **Solve the Equation or Inequality**: Use algebraic methods to solve the variable(s). Here's a basic approach:
    - Simplify both sides of the equation or inequality (combine like terms, distribute, etc.).
    - Use addition or subtraction to isolate the variable term.
    - Use multiplication or division to solve the variable.
6. **Check Your Solution:** Substitute your solution back into the original equation or inequality to verify it works.

Example:

Sarah has twice as many apples as Tom. Together, they have 18 apples. How many apples does each person have?

Solution:

1. Identify Variables: let x be the number of apples Tom has. Then Sarah has x apples.
2. Formulate Equation: Tom's apples + Sarah's apples= Total apples:
   $x + 2x = 18$.
3. Solve the Equation: Combine like terms:
   $3x = 18$
   Divide by 3: $x = 6$.
4. Check Your Solution: If Tom has 6 apples, Sarah has $2 \times 6 = 12$ apples, together, they have $6 + 12 = 18$ apples, which matches the problem statement.

Worksheets

Identify Expressions and Equations

✍ Identify whether each is an expression or an equation:

1) $x + 7 = 2$
2) $2y^2 + 9y + 5$
3) $x^2 - 5x$
4) $3z^3 - 3 = 0$
5) $0.7y + 2$
6) $-9 + 3x^2 = 6$
7) $9a + 8$
8) $\frac{2}{3}x - \frac{5}{10} + \frac{8}{9}x^2$
9) $48 - x^2 = -1$
10) $12 - y^2 + y$

One-Step Equations

✍ Solve.

1) $7 + x = -7$
2) $30 = x - 15$
3) $5x = 35$
4) $x + 15 = 20$
5) $12 = 16 + x$
6) $5 - x = -11$
7) $\frac{x}{3} = 0.2$
8) $36 - 12 = 4x$
9) $\frac{0.5}{5} = \frac{x}{10}$
10) $x - \frac{2}{5} = 0.75$

Two-Step Equations

✍ Solve.

1) $3 + 7x = -4$
2) $(-8)(3x - 4) = -16$
3) $2x + 12 = 16$
4) $8(-2x + 5) = -1$
5) $6(3 + x) = 42$
6) $\frac{x}{4} - 5 = 3$
7) $\frac{1}{4} = \frac{1}{2} + \frac{x}{4}$
8) $\frac{11 + x}{5} = (-6)$
9) $(-4) + \frac{x}{2} = (-14)$
10) $\frac{2x - 12}{8} = 6$

Multi-Step Equations

✍ Solve.

1) $-14(3 + x) = 14$
6) $18 - \frac{3}{5}x = -\frac{4}{5} - 5x$

2) $-3(2+x)=6$

3) $40(3+x)=40$

4) $-18=x+8x$

5) $3x+25=-2x-10$

7) $\frac{x-5}{2}=-5(-3-x)$

8) $-7-4x=6\frac{(4-x)}{3}$

9) $(-4)+\frac{x}{2}=(-14)$

10) $-3=\frac{-6x-9+3x}{2}$

Equation Diagram

Write an equation for each diagram:

1)

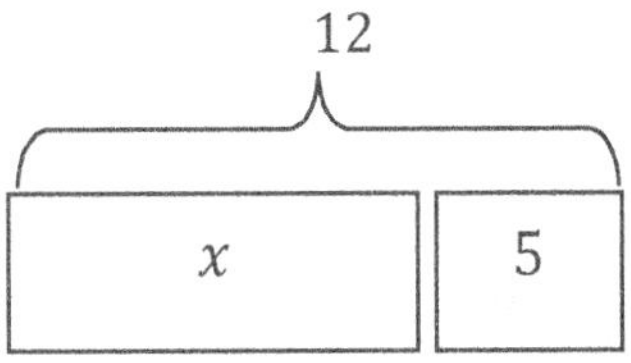

2)

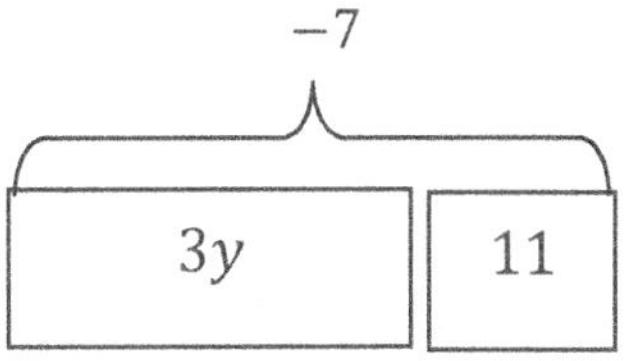

3)

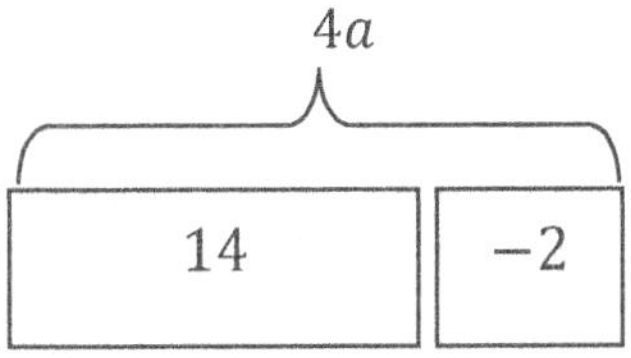

4)

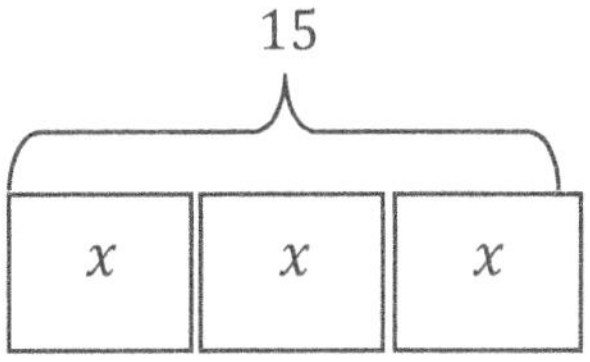

5)

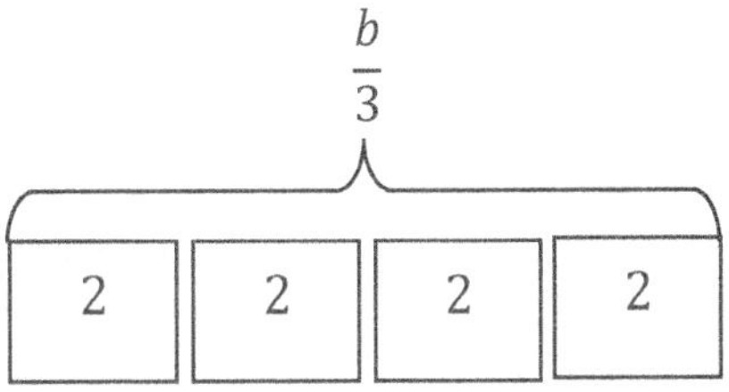

6)

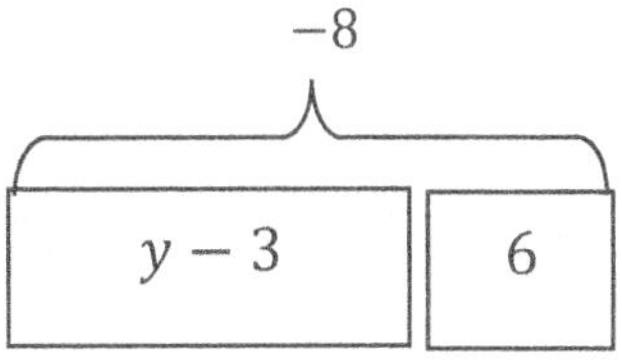

7)

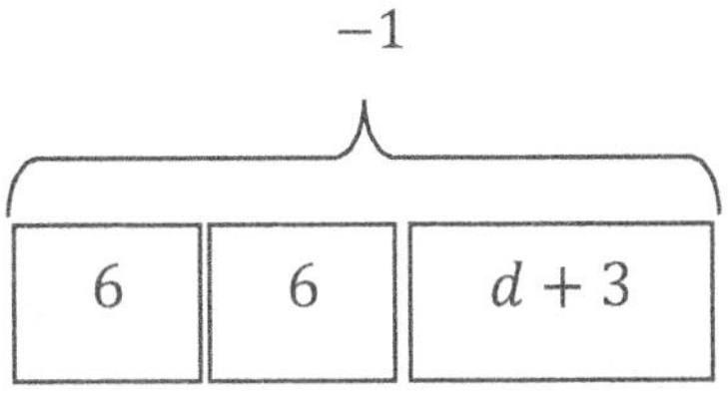

8)

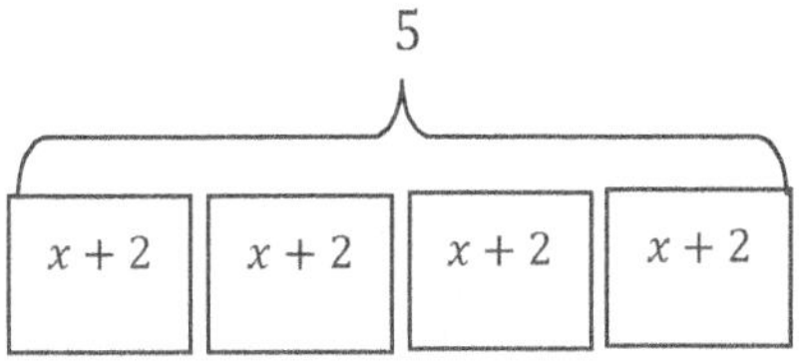

9)

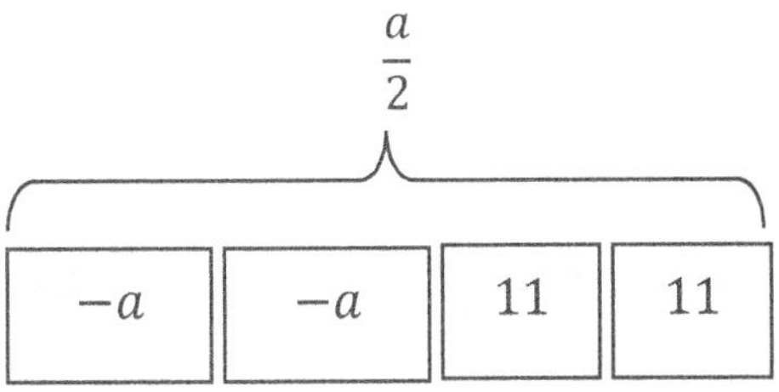

10)

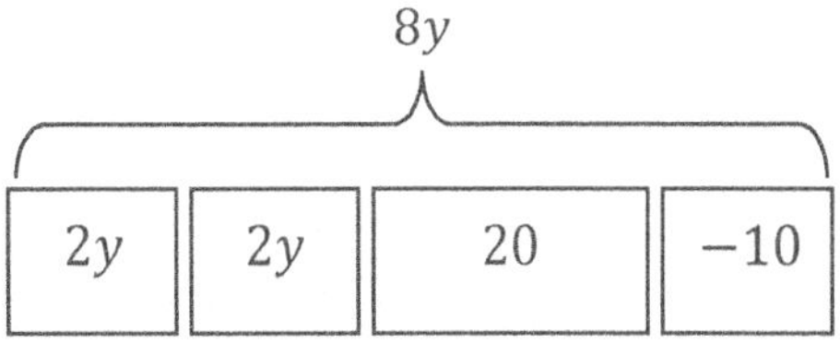

One-Step Inequalities

☙ Solve inequalities.

1) $4x < 24$
2) $x + 7 > 3$
3) $-y + 6 \geq 5$
4) $-2z \geq 8$
5) $-x - 5 \geq -7$
6) $y + \frac{3}{5} > 2$
7) $z \leq \frac{4}{9} - 0.2$
8) $x - 1 \geq -1$
9) $\frac{y}{-7} > \frac{4}{3}$
10) $3 - z \leq \frac{3}{10} + \frac{1}{4}$

Two-Step Inequalities

☙ Solve inequalities.

1) $2x + 5 \leq 13$
2) $5b - 2 \leq 18$
3) $7y - 3 \geq 10$
4) $-6z + 1 \geq 8$
5) $x - 5x \geq 16$
6) $\frac{x}{2} + 2 > -1$
7) $3z - 5 \leq 0.1$
8) $3(3x + 1) \geq 4$
9) $\frac{4z-4}{5} > 0.2$
10) $-x \leq \frac{x-3}{6}$

Multi-Step Inequalities

☙ Solve multi-step inequalities:

1) $4 - 6x < 36 + 2x$
2) $2(x - 3) + 5 \leq 11$
3) $3(y + 2) - 4 > 5$
4) $4(a - 1) + 2 \geq 3a + 7$
5) $6(c + 1) - 2c \leq 8 + 3c$
6) $\frac{3x-6}{2} < -\frac{1}{2}$
7) $-2y + \frac{3}{2} \leq 4y - 2$
8) $8(a + 3) - 2a \geq 6 + 4a - 5$
9) $\frac{-8z+20}{4} > \frac{2-z}{2}$
10) $\frac{5}{3}b - \frac{2}{4}(b - 3) < 4 + \frac{b}{6}$

Graphing Inequalities

☙ Graph the inequalities into the number line:

1) $x < 5$
2) $x \geq -2$
3) $x > \frac{3}{2}$
4) $x \leq 0$
5) $-x \geq -2$

Word Problems

✎ Solve following word problems by writing an equation or an inequality:

1) A movie ticket costs $8. If Jake buys t tickets and spends a total of $32, how many tickets did he buy?
2) The sum of a number and 7 is 12. What is the number?
3) Carlos has three times as many marbles as Ben. If Ben has m marbles and Carlos has 21 marbles, how many marbles does Ben have?
4) The length of a rectangle is 5 units more than its width. If the width is w and the perimeter of the rectangle is 30 units, what are the dimensions of the rectangle?
5) Anna has at most $30 to spend on books. Each book costs $6. How many books can she buy without exceeding her budget?
6) Jenny has $50 and she wants to buy some books that each cost $6. If she buys b books and still has at least $8 left, how many books can she buy?
7) A car rental company charges $20 per day plus $0.10 per mile driven. If Lisa rents a car for d days and drives m miles, her total cost is $100. If she drove 150 miles, for how many days did she rent the car?
8) A school needs to raise at least $500 for a field trip. If each student contributes $5, how many students need to contribute to reach or exceed the goal?
9) Alex spends $3 on a snack and the rest of his money on notebooks that each cost $2. If he has $15 and buys n notebooks, how many notebooks can he buy and still have at least $1 left?
10) The sum of three consecutive numbers is 36. What is the smallest number?

Answer of Worksheets

Identify Expressions and Equations

1) Equation
2) Expression
3) Expression
4) Equation
5) Expression
6) Equation
7) Expression
8) Expression
9) Equation
10) Expression

One-Step Equations

1) $x = -14$
2) $x = 45$
3) $x = 7$
4) $x = 5$
5) $x = -4$
6) $x = 16$
7) $x = 0.6$
8) $x = 6$
9) $x = 1$
10) $x = 1.15$

Two-Step Equations

1) $x = -1$
2) $x = 2$
3) $x = 2$
4) $x = \frac{41}{16}$
5) $x = 4$
6) $x = 32$
7) $x = -1$
8) $x = -41$
9) $x = -20$
10) $x = 30$

Multi-Step Equations

1) $x = -4$
2) $x = -4$
3) $x = -2$
4) $x = -2$
5) $x = -7$
6) $x = \frac{-47}{11}$
7) $x = \frac{-35}{9}$
8) $x = \frac{-15}{2}$
9) $x = -20$
10) $x = -1$

Equation Diagram

1) $x + 5 = 12$
2) $3y + 11 = -7$
3) $4a = 14 - 2$
4) $3x = 15$
5) $\frac{b}{3} = 8$
6) $y - 3 + 6 = -8$
7) $12 + d + 3 = -1$
8) $4(x + 2) = 5$
9) $-2a + 22 = \frac{a}{2}$
10) $4y + 20 - 10 = 8y$

One-Step Inequalities

1) $x < 6$
2) $x > -4$
3) $y \leq 1$
4) $z \leq -4$
5) $x \leq 2$
6) $y > \frac{7}{5}$
7) $z \leq \frac{11}{45}$
8) $x \geq 0$
9) $y < \frac{-28}{3}$
10) $z \geq \frac{49}{20}$

Two-Step Inequalities

1) $x \leq 4$
2) $b \leq 4$
3) $y \geq \frac{13}{7}$
4) $z \leq \frac{-7}{6}$
5) $x \leq -4$
6) $x > -6$
7) $Z \leq 1.7$
8) $x \geq \frac{1}{9}$
9) $z > 1.25$
10) $x \geq \frac{3}{7}$

Multi-Step Inequalities

1) $x > -4$
2) $x \leq 6$
3) $y > 1$
4) $a \geq 9$
5) $c \leq 2$
6) $x < \frac{5}{3}$
7) $y \geq \frac{7}{12}$
8) $a \geq -11.5$
9) $z < \frac{8}{3}$
10) $b < \frac{5}{2}$

Graphing Inequalities

1)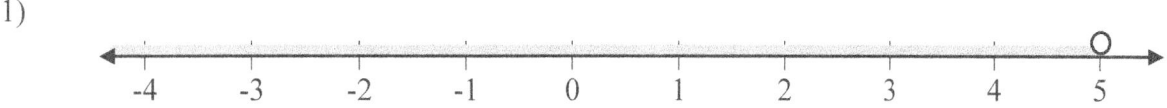

2)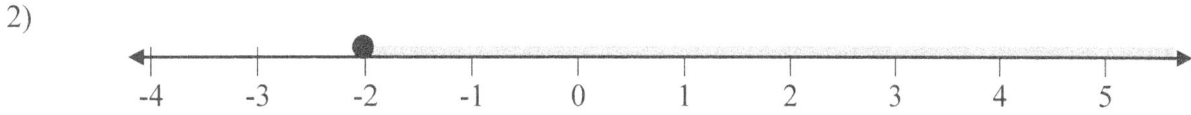

3)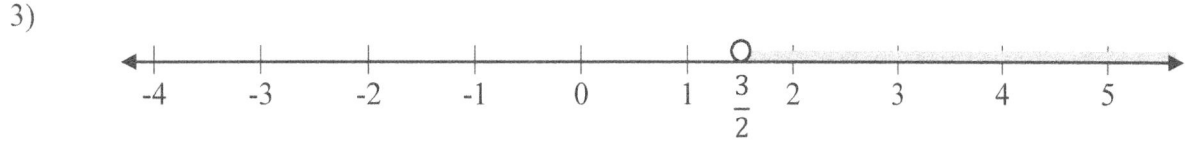

4)
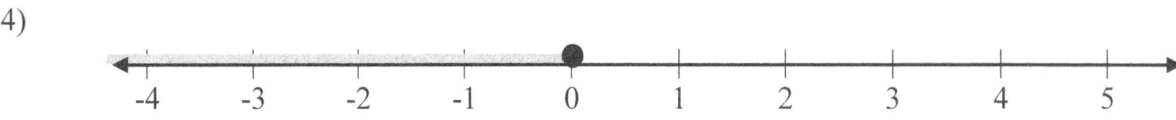

5)
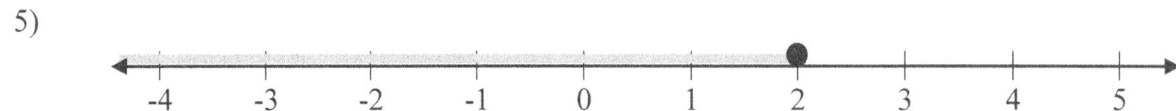

Word Problems

1) $8t = 32$, $t = 4$

2) $x + 7 = 12$, $x = 5$

3) $3m = 21$, $m = 7$

4) $30 = 2((w + 5) + w)$, $width = 5$ and $length = 10$

5) $6b \leq 30$, $b \leq 5$

6) $50 - 6b \geq 8$, $b \leq 7$

7) $100 = 20d + 0.1m$, $d = 4.25 \rightarrow d \approx 5$

8) $5n \geq 500$, $n \geq 100$

9) $12 - 2n \geq 1$, $n \leq 5.5$ so at least 5 notebooks

10) $x + (x + 1) + (x + 2) = 36$, $x = 11$

Chapter 11: Two Variable Equations

Topics that you will learn in this chapter:

- Describe Coordinate Plane
- Reflect a Point Over an Axis
- Distance Between Two Points
- Directions on Coordinate Plane
- Area and Perimeter of Shape on the Coordinate Plane
- Identify Independent and Dependent Variables
- Write an Equation from a Graph Using a Table
- Value of Two Variables Equations
- Word Problems
- Worksheet

Describe Coordinate Plane

A coordinate plane is a two-dimensional surface on which we can plot points, lines, and curves. It's used extensively in mathematics to visually represent relationships and solve problems involving two variables. Here's a breakdown of its components and features:

1. **Axes:**
   - $X-axis$: The horizontal axis.
   - $Y-axis$: The vertical axis.
2. **Origin:** The point where the $X-axis$ and $Y-axis$ intersect, designated as (0,0).
3. **Quadrants:** The plane is divided into four regions called quadrants:
   a) **First Quadrant:** Both x and y are positive.
   b) **Second Quadrant:** x is negative, and y is positive.
   c) **Third Quadrant**: Both x and y are negative.
   d) **Fourth Quadrant:** x is positive, and y is negative.
4. **Coordinates:** Points on the plane are represented by ordered pairs (x, y), where x is the value on the $X-axis$ and y is the value on the $Y-axis$.
5. **lotting Points:** To plot a point like $(3, 2)$, you move 3 units along the $X-axis$ and 2 units up the $Y-axis$.

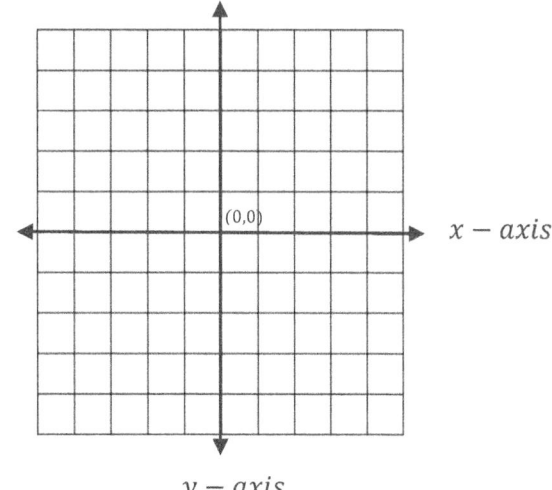

Example:

✱ Find points on the coordinate plane.
A(2,3)
B(−4,1)
C(−3,3)
D(4,−1)

Solution: To plot a point on coordinate plane, it is enough to move x (right for positive numbers and left for negative numbers) units along the $x-axis$ and y (right for positive numbers and left for negative numbers) units along the $y-axis$:

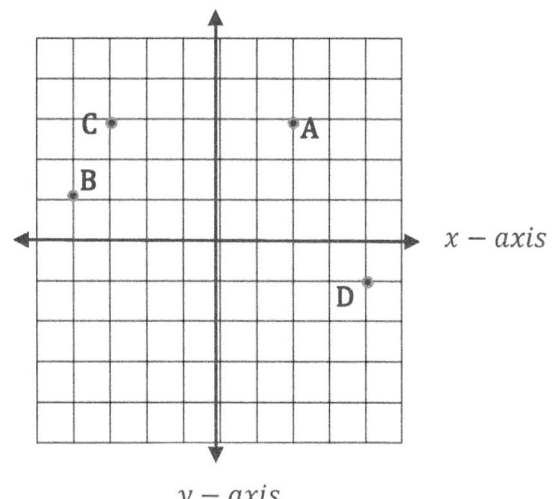

www.mathnotion.com

Directions on Coordinate Plane

Directions on a coordinate plane refer to the specific movements you make to navigate from one point to another on the grid. Understanding these directions is essential for plotting points, graphing lines, and solving geometric problems.

Positive and Negative Directions:

- **Positive X-Direction:** Moving to the right along the X-axis.
- **Negative X-Direction:** Moving to the left along the X-axis.
- **Positive Y-Direction:** Moving up along the Y-axis.
- **Negative Y-Direction:** Moving down along the Y-axis.

Moving from One Point to Another: To move from point $A\ (x_1, y_1)$ to point $B\ (x_2, y_2)$:

- Determine the change in the x-coordinates: $x_2 - x_1$
- Determine the change in the y-coordinates: $y_2 - y_1$

Direction Indicators:

- Moving right: $x_2 - x_1 > 0$
- Moving left: $x_2 - x_1 < 0$
- Moving up: $y_2 - y_1 > 0$
- Moving down: $y_2 - y_1 < 0$

Example:

✸ How do you move from point $A(2,3)$ to point $B(7,-1)$?

Solution:

1. Identify the Coordinates: point $A(2,3)$ and point $B(7,-1)$.
2. Calculating Horizontal Movement:
   - Difference in x-coordinates: $x_2 - x_1 = 7 - 2 = 5$
   - Move 5 units to the right.
3. Calculate Vertical Movement:
   - Difference in y-coordinates: $y_2 - y_1 = -1 - 3 = -4$
   - Move 4 units down.

Starting at $A(2,3)$, move 5 units right to $(7,3)$. Then, move 4 units down to $(7,-1)$. The point B is at $(7,-1)$.

Reflect a Point Over an Axis

Reflecting a point over an axis is a way to find its mirror image across that axis on the coordinate plane. Here's how you can reflect a point over the X-axis and the Y-axis:

Reflecting Over the X-Axis: When you reflect a point (x, y) over the $X-axis$, the X-coordinate stays the same, but the Y-coordinate changes sign.

- o **Original point:** (x, y)
- o **Reflected point:** $(x, -y)$

Reflecting Over the Y-Axis: When you reflect a point (x, y) over the $Y-axis$, the Y-coordinate stays the same, but the X-coordinate changes sign.

- o **Original point:** (x, y)
- o **Reflected point:** $(-x, y)$

Example

Reflect $A(2, 4)$ over the $X-axis$ and $B(-1, 3)$ over the $Y-axis$ and graph them on coordinate plane.

- **Solution:**
  - When you reflect a point $A(2, 4)$ over the $X-axis$, the X-coordinate (2) stays the same, but the Y-coordinate (4) changes sign. So, the reflected point will be: $A'(2, -4)$.
  - When you reflect a point $B(-1, 3)$ over the $Y-axis$, the Y-coordinate (3) stays the same, but the X-coordinate (-1) changes sign. So, the reflected point will be: $B'(1, 3)$.

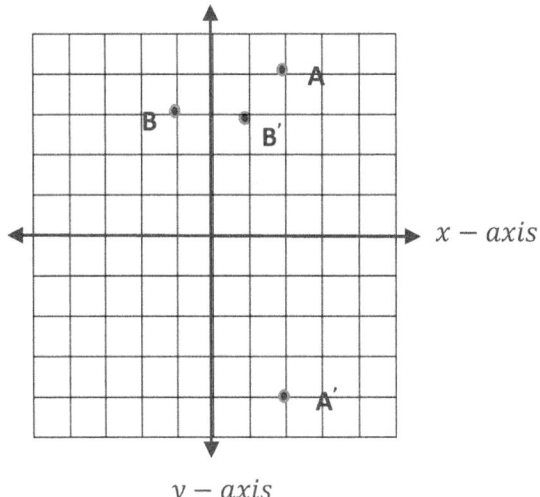

Distance Between Two Points

The distance between two points on a coordinate plane is equal to the length of the line segment that connects the two points.

Points with the same X-coordinate: If two points have the same X-coordinate, they lie on a vertical line. To find the distance between these two points, you can simply find the absolute difference between their y-coordinates:

- **Identify the Coordinates:** Let's say the two points are $A(x, y_1)$ and $B(x, y_2)$.
- **Calculate the Distance:** The distance between the two points is the absolute difference between their y-coordinates. Distance $= | y_2 - y_1 |$.

Points with the same Y-coordinate: If two points have the same Y-coordinate, they lie on a horizontal line. To find the distance between these two points, you can simply find the absolute difference between their X-coordinates:

- **Identify the Coordinates:** Let's say the two points are $A(x_1, y)$ and $B(x_2, y)$.
- **Calculate the Distance:** The distance between the two points is the absolute difference between their X-coordinates. Distance $= | x_2 - x_1 |$.

Examples:

1) Find the distance between two points: $A:(3, 1)$ and $B(3, -2)$.

   Solution:

   Since their X-coordinates are equal, the distance between the two points is the absolute difference between their y-coordinates.

   Distance $= | y_B - y_A | = |-2 - 1| = |-3| = 3$.

2) Find the distance between two points on the coordinate plane:
   Solution:
   - Identify the Coordinates: First, we need to determine the coordinates of the points. $A:(-4, 1)$ and $B(3, 1)$.
   - Calculate the Distance: The distance between the two points is the absolute difference between their X-coordinates.
   Distance $= | x_B - x_A | = |3 - (-4)| = |7| = 7$.

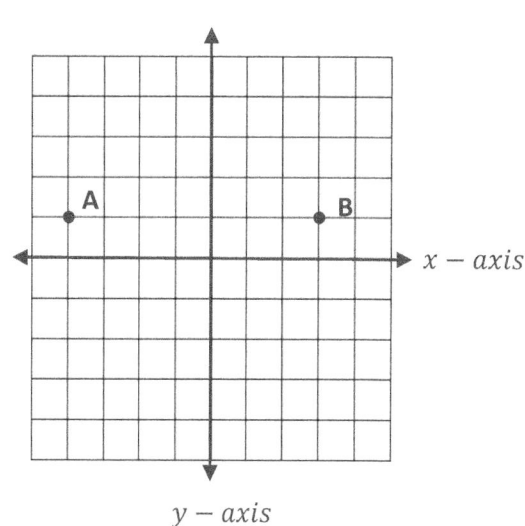

Identify Independent and Dependent Variables

Identifying independent and dependent variables is essential in understanding the relationship between two variables, especially in experiments and data analysis. The coordinate plane will help visualize the relationship between these variables more effectively:

1. **Independent Variable (X-axis):**
   - The independent variable is plotted on the horizontal axis (X-axis).
   - This variable is the one you control or change in an experiment.
2. **Dependent Variable (Y-axis):**
   - The dependent variable is plotted on the vertical axis (Y-axis).
   - This variable is the one you measure or observe to see the effect of the independent variable.

Steps to Plot the Graph:

1. **Draw the Axes:** Draw a horizontal line (X-axis) and a vertical line (Y-axis) intersecting at the origin (0,0).
2. **Label the Axes:** Label the horizontal axis as independent variable and the vertical axis as dependent variable.
3. **Plot the Points:** Plot the points based on the given data.
4. **Connect the points:** Draw a line through the points to show the trend.

Example:

Consider following table where x is the number of hours spent studying and y is the test score. Identify independent and dependent variables and plot the graph.

Study time	1	2	3	4
Test score	60	70	80	90

Solution:

Since the test score depends on the amount of study time, so y is dependent variable and x is independent variable.

For plotting the graph, we label (X-axis) as Study time and (Y-axis) as Test score and then plot the point below:

$(1,60), (2,70), (3,80), (4,90)$

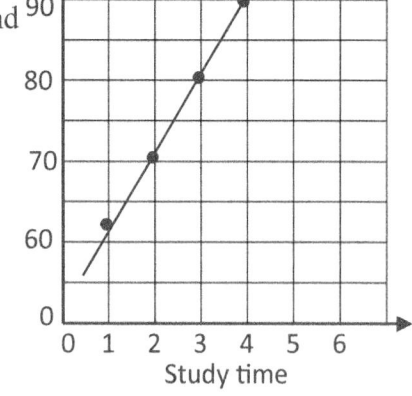

Area and Perimeter of Shape on the Coordinate Plane

To find the area and perimeter of a shape on the coordinate plane, follow these steps:

Finding the Perimeter:

1. **Identify the Vertices:** List the coordinates of all vertices of the shape.
2. **Calculate the Distance Between Adjacent Vertices:**
   - For points with the same x-coordinate, the distance between the two points is the absolute difference between their y-coordinates.
   - For points with the same y-coordinate, the distance between the two points is the absolute difference between their x-coordinates.
4. **Sum the Lengths of All Sides:** Add up the lengths of all sides to get the perimeter.

Finding the Area: The method for finding the area depends on the type of shape and its area formula, but in general, by counting each square unit on coordinate plane we can find the area.

Example:

Find the area and perimeter of a rectangle with following vertices:

$A: (1, 2), B: (1, 5), C: (6, 5), D: (6, 2)$

Solution:

1. **Perimeter:**
   - Points A and B have the same x-coordinate, so the distance between the two points is the absolute difference between their y-coordinates:
     Width= $|5 - 2| = |3| = 3$.
   - Points B and C have the same y-coordinate, so the distance between two points is the absolute difference between their x-coordinates:
     Length: $|6 - 1| = |5| = 5$.
   - Perimeter: The perimeter of rectangle is:
     $(Length + Width) \times 2 = (5 + 3) \times 2 = 16$ units.

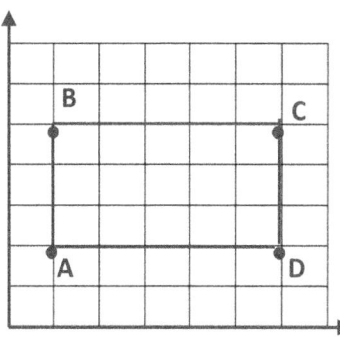

2. **Area:**
   The area of a rectangle is: $Length \times Width = 5 \times 3 = 15$ square units.

In general, we can find the perimeter by counting the segment units, and we can determine the area by counting the square units.

Write an equation from a Graph Using a Table

Writing an equation from a graph using a table involves identifying the relationship between the variables represented in the graph. Here's a simple step-by-step process to help you:

1. **Create a Table:** List the coordinates of several points from the graph in a table format.
2. **Determine the Pattern:** look for a pattern in how y changes as x changes by asking questions like: "When x goes up by 1, what happens to y?"
3. **Write a Simple Rule:**
   - Using the pattern, write a rule to connect x and y.
   - Think of the rules to find y if you know x.
4. **Write the Equation:**
   - Combine the rule into a simple equation format.
   - Use y for the output and x for the input.

Example:

Write an equation from graph using a table:

Solution:

Create a Table with Points from the Graph:

List these points as pairs of numbers in a table:

x	1	2	3	4
y	3	5	7	9

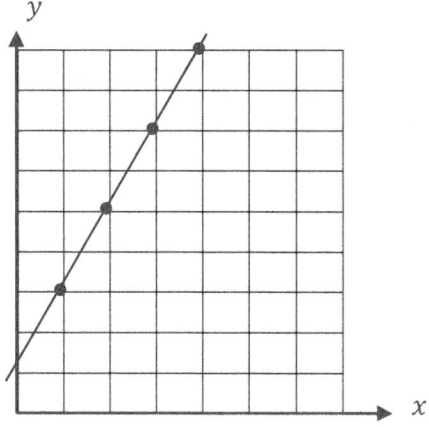

1. Find a Pattern: Look at how the y values change as the x values increase. Notice that as x goes up by 1, y goes up by 2.
2. Write a Simple Rule: Strat by guessing a rule: Think about doubling x and adding something to get y.
3. Testing the Rule:

   For $x = 1$: Doubling it gives $2 \times 1 = 2$. To get 3, we need to add 1.

   For $x = 2$: Doubling it gives $2 \times 2 = 4$. To get 5, we add 1.

   For $x = 3$: Doubling it gives $2 \times 3 = 6$. To get 7, we add 1.

   For $x = 4$: Doubling it gives $2 \times 4 = 8$. To get 9, we add 1.

   Rule: $y = 2 \times x + 1$

4. Write the Equation: Combine the rule into an equation format: Equation: $y = 2x + 1$.

www.mathnotion.com

Value of Two Variables Equations

Finding y When Given the Equation: To find y for a given equation, simply plug in different values for x and then calculate the result: here are the steps:

1. **Understand the Equation:** Look at the equation and understand the relation between x and y.
2. **Choose a Value for x**: Pick different values for x to see how y changes.
3. **Calculate y for each x:** Plug in different values for x and then calculate the result.

Finding x When Given the Equation: For different values of y, you'll need to solve for x. Here's how to do it step by step:

1. **Understand the Equation:** We need to rearrange this equation to solve for x.
2. **Isolate x:** Solve the equation to isolate x.
3. **Calculate x for Different Values of Y:** Substitute different y into equation and calculate x.

Examples:

1) Find the value of y using the equation $y = 5x - 1$ for $x = 1$.

   Solution:
   1. Understand the Equation: The equation we are working with is $y = 5x - 1$. This means that to find y, you multiply x by 5 and then subtract 1.
   2. Choose a Value for x: $x = 1$.
   3. Calculate y for each x. Plug $x = 1$ into the equation:

      $y = 5x - 1 = 5 \times 1 - 1 = 5 - 1 = 4$

      So, when $x = 1$, $y = 4$.

2) Find the value of x using the equation $y = -3x + 2$ for $y = -4$.

   Solution:
   1. Understand the Equation: The equation is $y = -3x + 2$ and we need to rearrange this equation to solve for x.
   2. Isolate x:
      - Subtract 2 from both sides of the equation to move +2 to the other side:
      $$y - 2 = -3x$$
      - Divide both sides by -3 to solve for x:
      $$x = \frac{y-2}{-3}$$
   3. Calculate x for Different Values of y: Substitute $y = -4$ into the new equation:
      $$x = \frac{y-2}{-3} = \frac{-4-2}{-3} = \frac{-6}{-3} = 2.$$
      So, when $y = -4$, $x = 2$.

Word Problems

Solving word problems involving two-variable equations can be approached systematically, here are steps to solve word problems with two-variable equations:

1. **Read the Problem Carefully:** Understand what the problem is asking. Identify key pieces of information and what you need to find.
2. **Define the Variables**: Assign variables to the quantities involved. For example, let x represent one unknown quantity and y represent another.
3. **Set Up the Equations**: Translate the word problem into mathematical equations using the defined variables.
4. **Solve for One Variable in Terms of the Other:** Rearrange the equation to express one variable in terms of the other.
5. **Substitute and Solve:** If additional information is provided, substitute the given values to find the variables.

Example:

You have $60 to spend on notebooks and pens. Each notebook costs $5 and each pen costs $2. Write an equation that represents the total cost if you buy n notebooks and p pens. In addition, if you decide to buy 8 notebooks, find the number of pens.

Solution:

1. Read the Problem: Understand that you need to represent the cost of notebooks and pens using an equation.
2. Define the Variables: Let' n be the number of notebooks. Let's be the number of pens.
3. Translate the Problem into an Equation: Each notebook costs $5, and each pen costs $2. The total cost should be $60.
   The equation is: $5n + 2p = 60$
4. Solve for One Variable in Terms of the Other: Rearrange the equation to solve for p:
   $5n + 2p = 60$
   $p = \frac{60-5n}{2}$
5. Substitute and Solve:
   $p = \frac{60-5n}{2} = \frac{60-5 \times 8}{2} = \frac{60-40}{2} = \frac{20}{2} = 10$
   So, if you buy 8 notebooks, you can buy 10 pens.

www.mathnotion.com

Worksheets

Describe Coordinate Plane:

✍Find the following points on coordinate plane and write which region each point is in:

1) (2,4)
2) (−4,5)
3) (1,−3)
4) (−1,0)
5) (−3,−4)

Directions on Coordinate Plane

✍How to move from point A to point B?

1) from $A(2,0)$ to $B(2,6)$
2) from $A(-1,5)$ to $B(2,-3)$
3) from $A(4,-1)$ to $B(3,-1)$
4) from $A(0,6)$ to $B(3,7)$
5) from $A(5,4)$ to $B(-4,-5)$
6) from $A(3,-8)$ to $B(0,1)$

✍Find the destination of A after following moves:

7) A(1,5), move 2 units to the right
8) A(2,−3), move 3 units up
9) A(−5,7), move 4 units to the left then 1 unit down
10) A (0,8), move 1 unit up then 3 units to the left

Reflect a Point Over an Axis

✍Reflect following point over the $X-axis$:

1) (−2,5)
2) (8,3)
3) (1,−9)
4) (0,3)
5) (4,0)

✍Reflect following point over the $Y-axis$:

6) (2,−1)
7) (−8,3)
8) (5,0)
9) (0,−3)
10) (−4,−4)

Distance Between Two Points

✍Find the distance between two points:

1) $A(-1,3)$ and $B(-1,7)$
2) $A(2,3)$ and $B(-2,3)$
3) $A(0,-5)$ and $B(0,9)$
4) $A(1.5,-4)$ and $B(-0.6,-4)$
5) $A(1,1.5)$ and $B\left(1,\frac{-1}{4}\right)$
6) $A\left(3\frac{2}{5},0\right)$ and $B(-2,0)$

✍ Find the distance between two points in each coordinate plane:

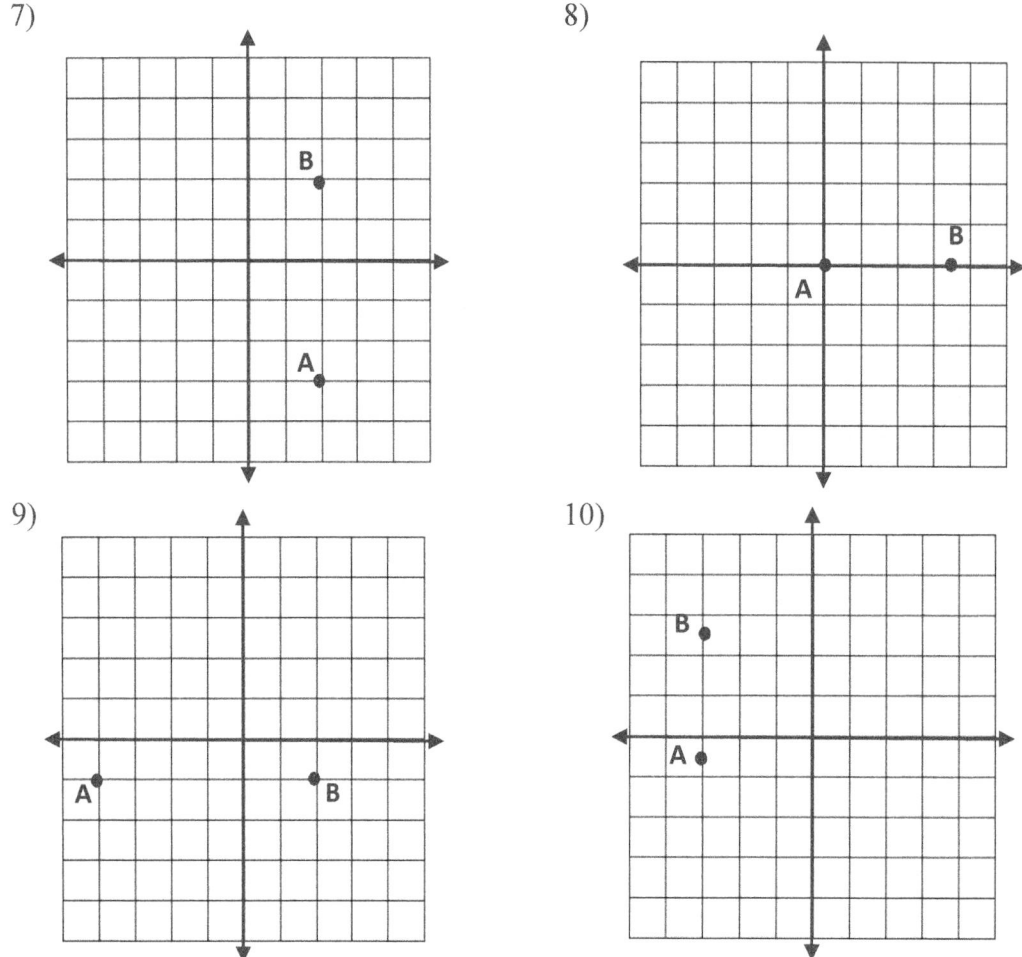

Identify Independent and Dependent Variables

✍ identify independent and dependent variables in following contexts:

1) You record the temperature at different times of the day. Determine the independent and dependent variables..
2) You measure the height of a plant every week for 6 weeks. Identify the variables.
3) You track the distance you run each day over a week and the time it takes. Identify the variables.
4) You monitor your heart rate before and after exercise. Determine the variables.
5) You keep track of how many pages you read each day and the total reading time. Identify the variables.

Area and Perimeter of Shape on the Coordinate Plane

📘 Find area and perimeter of shapes with following vertices:

1) $(0,0), (2,0), (2,2), (0,2)$
2) $(1,1), (6,1), (1,4), (6,4)$
3) $(0,-4), (-3,-4), (-3,0), (0,0)$
4) $(-4,-4), (-4,4), (4,4), (4,-4)$
5) $(0,0), (0,-3), (-4,0)$ (calculate the area only)
6) $(0,0), (3,4), (6,0), (9,4)$ (calculate the area only)

📘 Find area of all shapes and perimeter of first shape on coordinate plane:

7)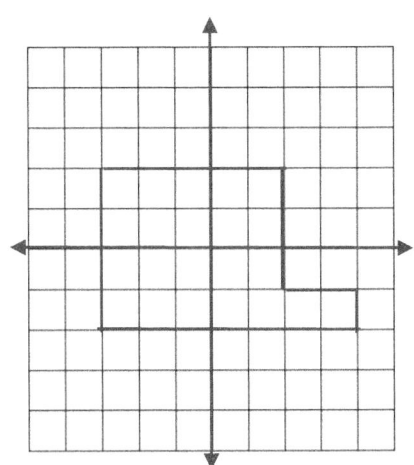

8)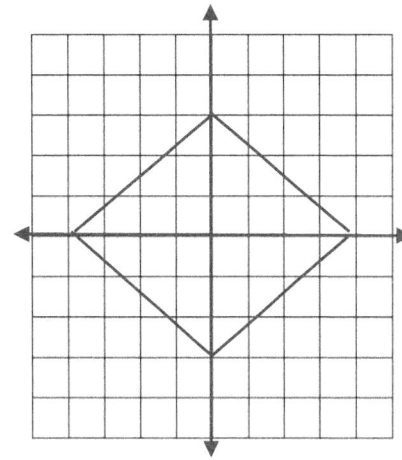

9)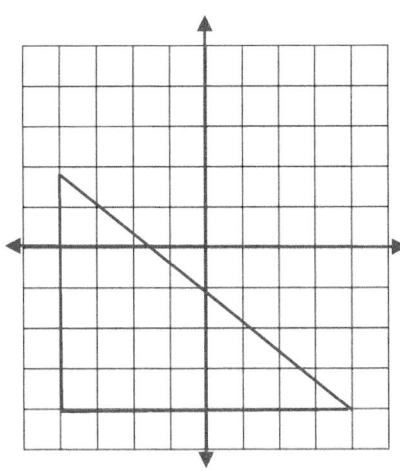

10)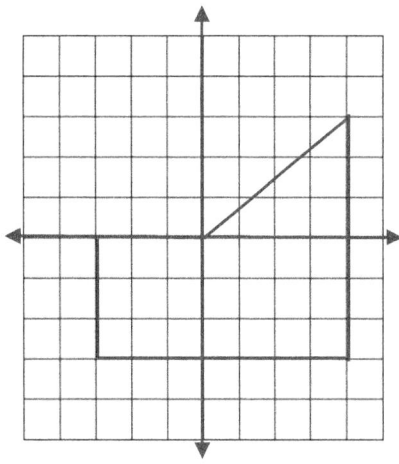

Value of Two Variables Equations

📘 Find the value of y using the equations below:

1) $y = 2x - 2$ and $x = 3$
2) $y = -4x + 1$ and $x = 0$
3) $y = \frac{1}{2}x - 3$ and $x = -4$
4) $y = 0.3x + 5$ and $x = 10$

www.mathnotion.com

5) $y = \frac{2}{3}x - \frac{1}{4}$ and $x = 6$

✍ Find the value of *x* using the equations below:

6) $y = -x + 1$ and $y = 5$
7) $y = 3x - 4$ and $y = 1$
8) $y = -x + 7$ and $y = -2$
9) $y = \frac{1}{4}x - 3$ and $y = 0$
10) $y = 0.5x + 2$ and $y = 10$

Write an Equation from a Graph Using a Table:

✍ Write an equation from graphs:

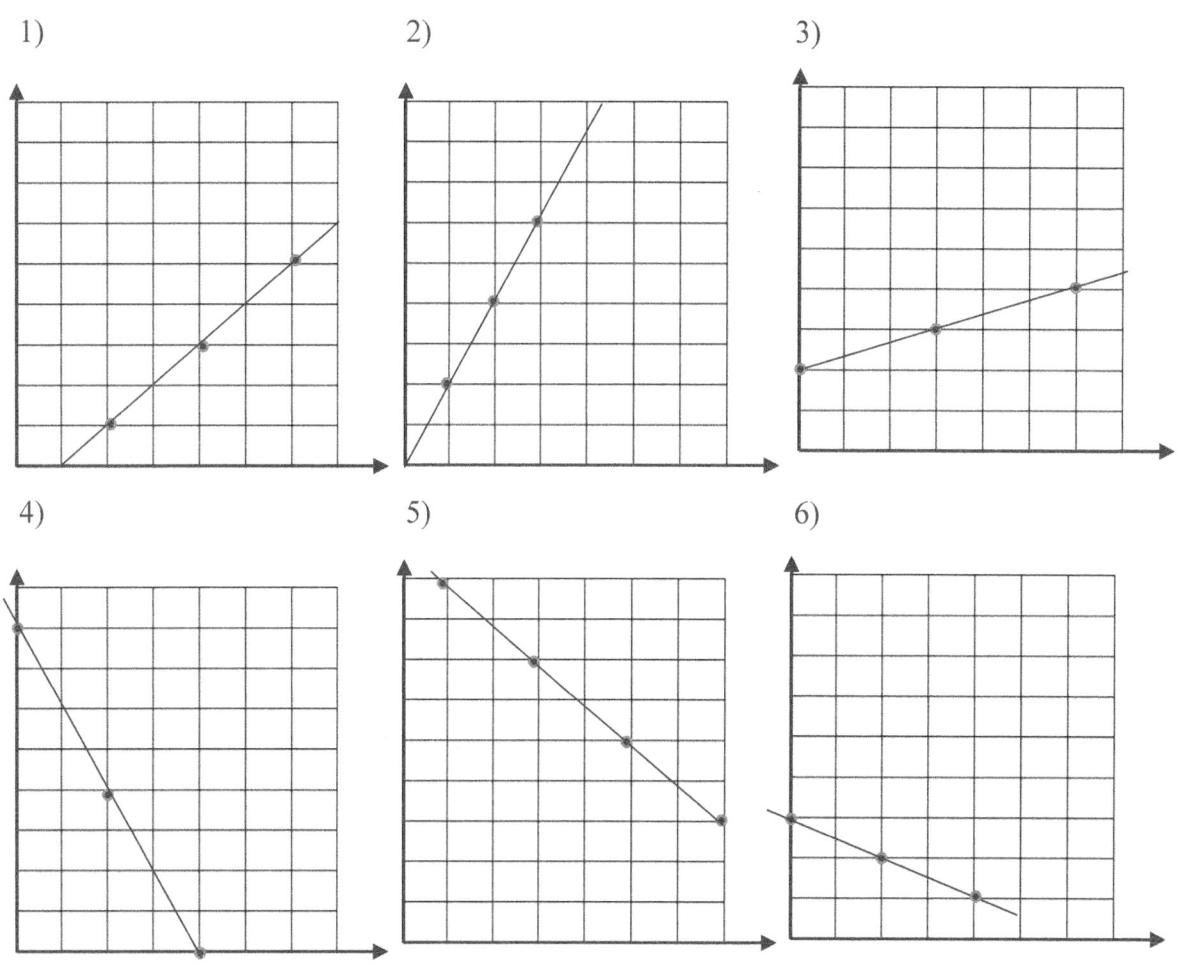

1)
2)
3)
4)
5)
6)

Word Problems

Do following word problems by writing an equation with two variables:

1) Sarah bought some apples and bananas. Each apple costs $1 and each banana costs $2. She spent a total of $10. Write an equation and find how many apples and bananas she could have bought.

2) John has twice as many pencils as erasers. Together, he has 12 items. Write an equation to represent this situation.

3) A teacher buys a total of 30 books and markers. She buys 4 books for every 5 markers. How many books did she buy?

4) In the garden, there are roses and tulips. The total number of flowers is 20. If the number of roses is 3 times the number of tulips, write an equation to represent this situation. Then, find the numbers of roses and tulips.

5) If a lemonade stand sells 2 cups of lemonade for every 3 cups of iced tea, and they sell a total of 18 cups of lemonade, how many cups of iced tea did they sell?

6) The length of a rectangle is 3 times its width, if the perimeter of this rectangle is 56 units, find its dimensions.

7) The sum of the ages of Peter and his brother is 21 years. If Peter is 5 years older than his brother, find the age of each one.

Answer of Worksheets

Describe Coordinate Plane:
1) First quadrant
2) Second quardrant
3) Fourth quadrant
4) On X-axis
5) Third quadrant

Directions on Coordinate Plane
1) 6 units up
2) 3 units to the right and 8 units down
3) 1 unit to the left
4) 3 units to the right and 1 unit up
5) 9 units to the left and 9 units down
6) 3 units to the left and 9 units up
7) (3,5)
8) (2,0)
9) (−9,6)
10) (−3,9)

Reflect a Point Over an Axis
1) (−2,−5)
2) (8,−3)
3) (1,9)
4) (0,−3)
5) (4,0)
6) (−2,−1)
7) (8,3)
8) (−5,0)
9) (0,−3)
10) (4,−4)

Distance Between Two Points
1) 4 units
2) 4 units
3) 14 units
4) 2.1 units
5) 1.75 units
6) 5.4 units
7) 5 units
8) 3.5 units
9) 6 units
10) 3 units

Area and Perimeter of Shape on the Coordinate Plane
1) Perimeter: 8 *units*, Area: 4 *square units*
2) Perimeter: 16 *units*, Area: 15 *square units*
3) Perimeter: 14 *units*, Area: 12 *square units*
4) Perimeter: 32 *units*, Area: 64 *square units*
5) Area: 6 *square units*
6) Area: 24 *square units*
7) Perimeter: 22 *units*, Area: 22 *square units*
8) Area: 24 *square units*
9) Area: 24 *square units*
10) *Area*: 27 *square units*

Identify Independent and Dependent Variables

1) Independent Variable: Time of the day (hours)
   Dependent Variable: Temperature (°C)
2) Independent Variable: Number of weeks
   Dependent Variable: Height of the plant (cm)
3) Independent Variable: Distance run (km)
   Dependent Variable: Time taken (minutes)
4) Independent Variable: Time (minutes)
   Dependent Variable: Heart rate (beats per minute)
5) Independent Variable: Number of pages read
   Dependent Variable: Reading time (minutes)

Write an Equation from a Graph Using a Table:

1) $y = x - 1$
2) $y = 2x$
3) $y = \frac{1}{3}x + 2$
4) $y = -2x + 8$
5) $y = -x + 10$
6) $y = -\frac{1}{2}x + 3$

Value of Two Variables Equations

1) $y = 4$
2) $y = 1$
3) $y = -5$
4) $y = 8$
5) $y = 3\frac{3}{4}$
6) $x = -4$
7) $x = \frac{5}{3}$
8) $x = 9$
9) $x = 12$
10) $x = 16$

Word Problems

1) a represent the number of apples and b represent the number of bananas, $a + 2b = 10$
2) p represent the number of pencils and e represent the number of erasers, $p + e = 12$ and $p = 2e$, so $2e + e = 12$
3) b represent the number of books and m represent the number of markers, $b + m = 30$ and $b = \frac{4}{5}m$, she buys 12 books.
4) r represent the number of roses and t represent the number of tulips, $r + t = 20$ and $r = 3t$ so, $3t + t = 20$, number of roses: 15, number of tulips: 5.
5) l represent the number of cups of lemonade sold and t represent the number of cups of iced tea sold. $l = \frac{2}{3}t$, the lemonade stand sold 27 cups of iced tea.
6) w represent the width of the rectangle and l represent the length of the rectangle. $2l + 2w = 56$ and $l = 3w$. Width: 7 units and Length: 21 units
7) p represent Peter's age and b represent his brother's age. $p + b = 21$ and $p = b + 5$. Peter's age: 13 years and His brother's age: 8 years.

Chapter 12: Geometry and Solid Figures

Topics that you will learn in this chapter:

Introduction to Geometry

- Lines, Segments, and Angles
- Complementary and Supplementary Angles
- Symmetry and Congruence
- Classifying Polygons
- Classifying Triangles
- Pythagorean Relationship

Area and Perimeter

- Rectangles
- Parallelograms
- Triangles
- Trapezoids
- Circles
- Area of Compound Shapes

Surface Area and Volume

- Cubes
- Rectangular Prisms
- Triangular Prisms
- Cylinder
- Worksheet

Lines, Segments, and Angles

Lines: A line is a straight one-dimensional figure that has no thickness and extends infinitely in both directions.

Properties: Lines are often named using lowercase letters or by any two points that lie on the line (e.g., line $\overleftrightarrow{x}$ or line $\overleftrightarrow{AB}$).

Segments: A segment is a part of a line that is bounded by two distinct end points.

Properties: It has a definite beginning and end (e.g., segment $\overline{AB}$ where A and B are the endpoints).

Angles: An angle is formed by two rays (the sides of the angle) that share a common endpoint (the vertex of the angle).

Properties: Angles are measured in degrees (°) or radians. Types of angles include:

- Acute Angle: Less than 90°
- Right Angle: Exactly 90°
- Obtuse Angle: Greater than 90° but less than 180°
- Straight Angle: Exactly 180°

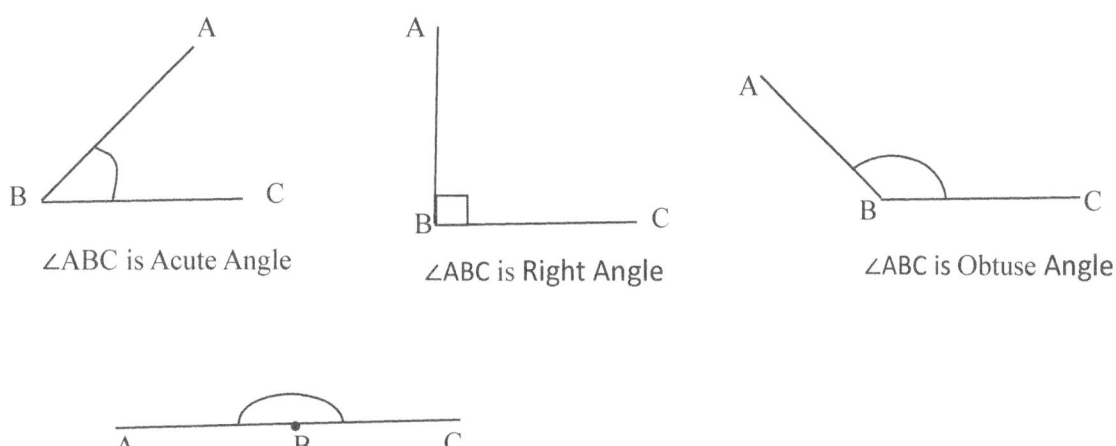

Complementary and Supplementary Angles

Complementary Angles: Two angles are called complementary when their measures add to 90 degrees. They do not have to be adjacent or form a straight line.

Supplementary Angles: Two angles are called supplementary when their measures add up to 180 degrees. Like complementary angles, supplementary angles do not have to be adjacent, but they often appear as angles on a straight line.

To find the complementary or supplementary angle of a given angle, you need to use the definitions and apply simple subtraction. Here's how:

Finding the Complementary and Supplementary of angles:

- To find the complementary angle, subtract the given angle from 90°.
- To find the supplementary angle, Subtract the given angle from 180°.

Examples:

1) Find the complementary of given angles:
   a) 50° b) 72°

   Solution:

   To find the complementary angle, subtract the given angle from 90°:

   a) 90° − 50° = 40° b) 90° − 72° = 18°

2) Find the Supplementary of given angles:
   a) 110° b) 95°

   Solution:

   To find the supplementary angle, subtract the given angle from 180°:

   a) 180° − 110° = 70° b) 180° − 95° = 85°

3) Based on the given ∠ABC, find the angle x:

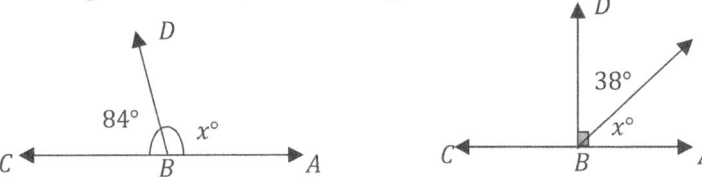

   Solution:

   According to the left shape, two angles ∠CBD and ∠DBA are supplementary so:

   $x = 180° - 84° = 96°$

   And about the right shape, two angles ∠DBE and ∠EBA are complementary so:

   $x = 90° - 38° = 52°$

Symmetry and Congruence

Symmetry: Symmetry refers to a property of a shape or object where one part of the shape is a mirror image or an identical counterpart of another part. There are different types of symmetrical, such as:

- **Line (or Reflection) Symmetry**: A shape has line symmetry if one half is a mirror image of the other half when divided by a line (called the line of symmetry).
- **Rotational Symmetry:** A shape has rotational symmetry if it can be rotated at a central point and still looks the same at certain angles.
- **Point Symmetry:** A shape has point symmetry if every part of the shape has an identical counterpart at an equal distance from a central point.

Congruence: Congruence refers to the idea that two shapes or figures are identical in size and shape, though they may be positioned differently. Congruent figures have corresponding sides and angles that are equal in length and measure.

- The two shapes are congruent if they can be made to coincide exactly by a combination of translation (sliding), rotation (turning), and reflection (flipping).

Examples:

1) Which type of symmetry do the following shapes have?

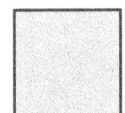  

Solution:

Square: A square has all types of symmetry, it has 4 lines of reflectional symmetry, it can be rotated 90°, 180° and 270° about its central point and finally has point symmetry.

Regular Pentagon: A regular pentagon has rotational symmetry of 72° (360 ÷ 5)144°and 216°. In addition, it has 5 lines of reflectional symmetry.

Parallelogram: A parallelogram just has point symmetry.

2) Which of following triangles are congruent?

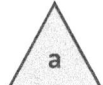 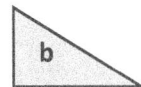 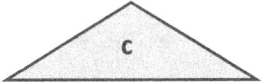 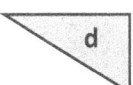 

Solution:

Shapes *a* and *d* are congruent because they are identical in size and shape, and they can be made to coincide by rotation (180°).

Classifying Polygons

Polygons can be classified based on several characteristics such as the number of sides, regularity, and convexity. Here are some most important kinds of classifying:

1. **By Number of Sides**: Polygons are named based on the number of sides they have:

Triangles: 3 sides	Quadrilateral: 4 sides	Pentagon: 5 sides	Hexagon: 6 sides	Heptagon: 7 sides	Octagons: 8 sides	Nonagon: 9 sides	Decagon: 10 sides

2. **By Regularity:**
   - **Regular Polygon:** All sides and angles are equal (e.g., an equilateral triangle, a square).
   - **Irregular Polygon**: Sides and angles are not all equal.
3. **By Convexity:**
   - **Convex Polygon**: All interior angles are less than 180°, and no vertices point inward.
   - **Concave Polygon:** At least one interior angle is greater than 180°, creating an indentation.

Examples:

1) Which polygon is regular, and which is irregular?

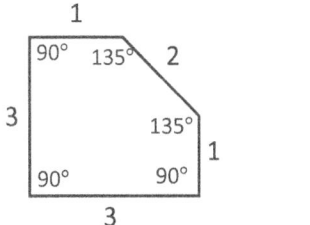 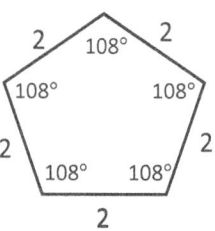

Solution:

A regular polygon has equal sides and angles, so the right shape is a regular polygon.

2) Which shape is a concave polygon?

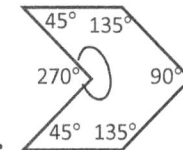 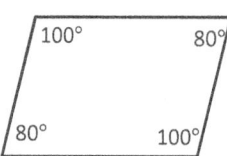

Solution:

A concave polygon has at least one interior angle greater than 180°, so the left shape is a concave polygon.

Classifying Triangles

Triangles can be classified in two main ways: by their sides and by their angles.

1. **Classification by Sides**

    - **Equilateral Triangle:** All three sides are equal in length, and all three angles are equal (each measuring 60°).
    - **Isosceles Triangle:** Two sides are equal in length, and the angles opposite those sides are equal.
    - **Scalene Triangle:** All three sides have different lengths, and all three angles are different.

2. **Classification by Angles**

    - **Acute Triangle**: All three angles are less than 90°.
    - **Right Triangle**: One angle is exactly 90° (a right angle) and the side opposite the 90° angle is called the hypotenuse.
    - **Obtuse Triangle**: One angle is greater than 90° but less than 180°.

    ☑ The sum of the angles of all triangles is equal to 180°.

Examples:

1) Which kind of triangle are following triangles? (Equilateral, Isosceles or Scalene)

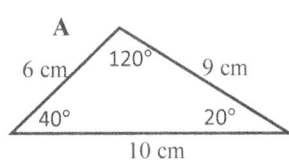 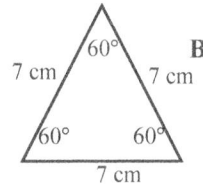 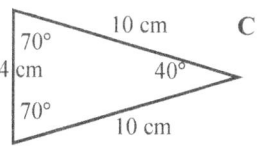

Solution:

According to their sides and their angles, triangle A is scalene, triangle B is equilateral, and triangle C is isosceles.

2) Which kind of triangle are following triangles? (acute, right or obtuse)

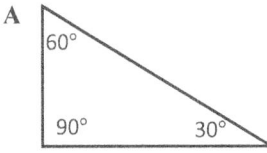 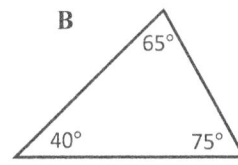 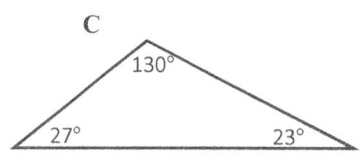

Solution:

According to their angles' sizes, triangle A is right, triangle B is acute, and triangle c is obtuse.

Pythagorean Relationship

The Pythagorean Relationship, also known as the Pythagorean Theorem, is a fundamental principle in geometry. It relates the lengths of the sides of a right-angled triangle. The theorem is named after the ancient Greek mathematician Pythagoras.

The Pythagorean Theorem

The theorem states: For any right-angled triangle, the square of the length of the hypotenuse (the side opposite the right angle) is equal to the sum of the squares of the lengths of the other two sides.

Mathematically, it is expressed as: $a^2 + b^2 = c^2$

where:

- a and b are the lengths of the legs (the two shorter sides of the triangle),
- c is the length of the hypotenuse (the longest side of the triangle).

Examples:

1) Find x and y in following right angles:

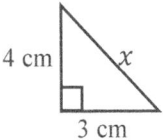
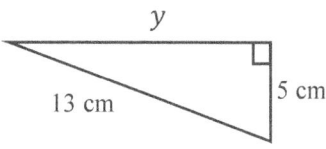

Solution:

Based on the Pythagorean Relationship for right triangles we have:

- $4^2 + 3^2 = x^2$; $16 + 9 = x^2$
  $25 = x^2$ now take the square root of both sides to find the hypotenuse:
  $x = \sqrt{25} = 5\ cm$.
- $y^2 + 5^2 = 13^2$; $y^2 + 25 = 169$
  then $y^2 = 169 - 25 = 144$, so like the first part, $y = \sqrt{144} = 12$.

2) Find the value of x:

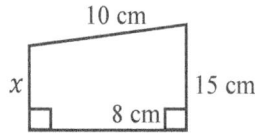

Solution: since we have a right triangle above the shape, we can consider Pythagorean Relationship: $8^2 + y^2 = 10^2$, $y^2 = 10^2 - 8^2 = 100 - 64 = 36$ and $y = \sqrt{36} = 6$

so, $x = 15 - 6 = 9\ cm$.

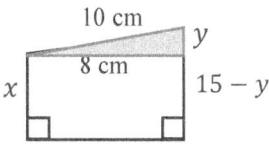

Parallelograms

Parallelograms are a type of quadrilateral (a four-sided polygon) with some distinct properties. They are essential shapes in geometry with unique characteristics.

Characteristics of Parallelograms:

- **Opposite Sides are Parallel:** Both pairs of opposite sides are parallel to each other.
- **Opposite Sides are Equal:** The lengths of opposite sides are equal.
- **Opposite Angles are Equal:** The measures of opposite angles are equal.
- **Consecutive Angles are Supplementary:** The sum of the measures of two consecutive angles is 180°.

Properties of Parallelograms:

- **Diagonals:** The diagonals of a parallelogram bisect each other, which means they cut each other into two equal parts.
- **Area:** The area A of a parallelogram can be calculated using the formula:

$$A = base \times height$$

- **Perimeter:** The perimeter P of a parallelogram can be calculated using the formula:

$$P = 2 \times (length + width)$$

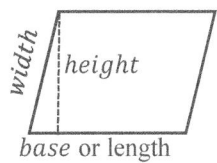

Examples:

1) Find the missing values:

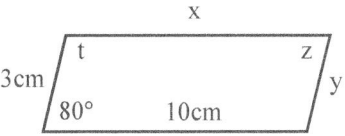

   Solution:
   - The lengths of opposite sides are equal, so, y=3 cm and x=10 cm.
   - The measures of opposite angles are equal, so, z=80°
   - The sum of the measures of two consecutive angles is 180°, so, t=180°-80°=100°

2) Calculate the area and perimeter the following shapes:

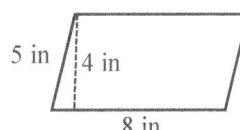

   Solution:
   - $A = base \times height = 8 \times 4 = 32$ square in
   - $P = 2 \times (length + width) = 2 \times (5 + 8) = 2 \times 13 = 26$ in

Rectangles

A rectangle is a parallelogram with all angles equal to 90°. They are essential shapes in geometry with unique characteristics.

Characteristics of Parallelograms:

- **Opposite Sides are Equal:** The lengths of opposite sides are equal.
- **All Angles are Right Angles:** Each of the four angles in a rectangle is 90°.
- **Parallel Sides:** The opposite sides are parallel to each other.

Properties of Parallelograms:

- **Diagonals:** The diagonals of a rectangle are equal in length, and they bisect each other.
- **Area:** The area A of a rectangle can be calculated using the formula:
$$A = length \times width$$
- **Perimeter:** The perimeter P of a rectangle can be calculated using the formula:
$$P = 2 \times (length + width)$$

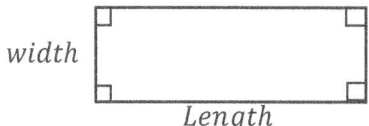

Examples:

1) A rectangle has a length of 10 cm and a width of 6 cm. What is its area and perimeter?
   Solution:
   - The area of a rectangle is given by the formula:
   $$A = length \times width = 10 \times 6 = 60\ cm^2$$
   - The perimeter of a rectangle is given by the formula:
   $$P = 2 \times (length + width) = 2 \times (10 + 6) = 2 \times 16 = 32\ cm$$

2) A rectangle has a width of 4 meters and an area of 32 square meters, find the length and perimeter of rectangle.
   Solution:
   - Find the Length: The area of a rectangle is given by formula:
   $Area = length \times width$
   $32 = length \times 4$. For the length we must divide 32 by the 4:
   length $= 32 \div 4 = 8$ meters.
   - Find the perimeter: The perimeter of a rectangle is given by the formula:
   $Perimeter = 2 \times (length + width) = 2 \times (8 + 4) = 2 \times 12 = 24\ meters.$

Triangles

Perimeter of a Triangle:

The perimeter of a triangle is the sum of the lengths of its three sides. If a triangle has sides a, b, and c, the perimeter P is given by:

$$P = a + b + c$$

Area of a Triangle:

The area of a triangle can be calculated by the base (b) and height (h) of the triangle:

$$A = \frac{1}{2} \times b \times h$$

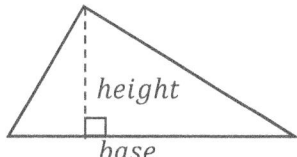

Examples:

1) Calculate the perimeter of a triangle with sides $a = 5$ cm, $b = 6$ cm and $c = 7$ cm.

   Solution: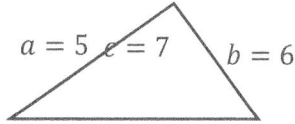

   According to the perimeter formula:

   $P = a + b + c = 5 + 6 + 7 = 18$ cm.

2) Calculate the area of a triangle with base = 4 inches and height = 5 inches:
   Solution:

   According to the area formula:

   $A = \frac{1}{2} \times b \times h = \frac{1}{2} \times 4 \times 5 = \frac{1}{2} \times 20 = 10$ square inches.

3) Calculate the perimeter of the triangle based on the figure below:
   Solution:

   For calculating the perimeter, we need the other side (we call it x) too, since this triangle is a right triangle, so we use the Pythagorean Relationship: $6^2 + 8^2 = x^2$ then $36 + 64 = 100 = x^2$ so, $x = 5$ in

   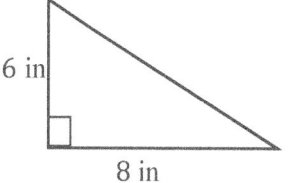

   and the perimeter is: $6 + 8 + 10 = 24$ in.

Trapezoids

Trapezoids are a type of quadrilateral (a four-sided polygon) with at least one pair of parallel sides.

Characteristics of Trapezoids:

- **Parallel Sides:** At least one pair of opposite sides is parallel. These are called the bases.
- **Non-parallel Sides:** The other two sides are called the legs, which are not necessarily equal or parallel.

Properties of Trapezoids:

- **Area:** The area A of a trapezoid can be calculated using the formula:

$$A = \frac{1}{2} \times (base_1 + base_2) \times height$$

Where $base_1$ and $base_2$ are the lengths of the two parallel sides, and the height is the perpendicular distance between these bases.

- **Perimeter:** The perimeter P of a trapezoid is the sum of the lengths of all its sides:

$$P = base_1 + base_2 + leg_1 + leg_2$$

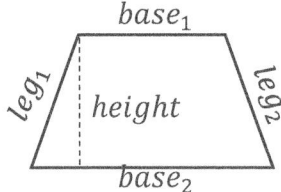

- The legs of a trapezoid are not necessarily equal.

Example:

Calculate the area and the perimeter of the following trapezoid:

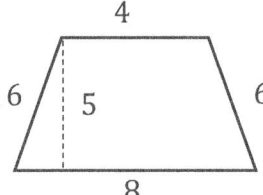

Solution:

- The area A of a trapezoid can be calculated using the formula:

$A = \frac{1}{2} \times (base_1 + base_2) \times height = \frac{1}{2} \times (4 + 8) \times 5 = 30$ The perimeter P of a trapezoid is the sum of the lengths of all its sides:

$$P = base_1 + base_2 + leg_1 + leg_2 = 4 + 8 + 6 + 6 = 24$$

Circles

A circle is a set of all points in a plane that are equidistant from a fixed point called the center. The fixed distance from the center to any point on the circle is called the radius.

Key Components

- **Center:** The fixed point from which all points on the circle are equidistant.
- **Radius:** The distance from the center of the circle to any point on the circle.
- **Diameter:** The distance across the circle, passing through the center. It's twice the length of the radius: $Diameter = 2 \times Radius$

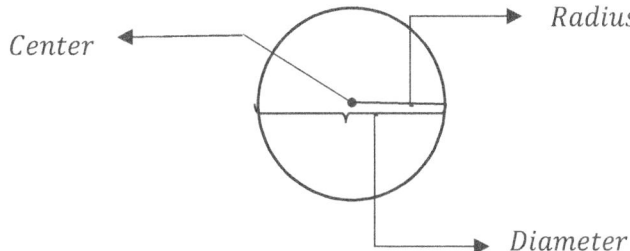

- **Circumference:** The distance around the circle. It can be calculated using the formula:
$$Circumference = 2\pi \times Radius = \pi \times Diameter$$
- **Area:** The amount of space enclosed by the circle. It can be calculated using the formula:
$$Area = \pi \times Radius^2$$
  ☑ The ratio of the circumference to the diameter is always π (approximately 3.14).

Examples:

1) A circle has a radius of 7 cm. Find its area and circumference.
   Solution:
   - Circumference: $2\pi \times Radius = 2\pi \times 7 = 14\pi \approx 43.96$ cm.
   - Area: $\pi \times Radius^2 = \pi \times 7^2 = \pi \times 49 \approx 153.86$ square cm.

2) A circle has a diameter of 10 m. Find its area and circumference.
   Solution:
   1. Radius: $Radius = \frac{Diameter}{2} = \frac{10}{2} = 5$.
   2. Circumference: $2\pi \times Radius = 2\pi \times 5 = 10\pi \approx 31.42$ m.
   3. Area: $\pi \times Radius^2 = \pi \times 5^2 = \pi \times 25 \approx 78.54$ square cm.

Area of Compound Shapes

The area of compound shapes (also known as composite shapes) involves finding the total area of a shape that is composed of two or more simple geometric shapes. To find the area of a compound shape, you typically break it down into simpler shapes, calculate the area of each, and then combine those areas.

Steps to Find the Area of Compound Shapes:

1. **Identify and Separate Simple Shapes:** Divide the compound shape into simpler shapes such as rectangles, triangles, circles, etc.

2. **Calculate the Area of Each Simple Shape:** Use the appropriate formula for each simple shape:
   - **Rectangle**: Area= $length \times width$
   - **Triangle**: Area= $\frac{1}{2} \times base \times height$
   - **Circle**: Area= $\pi \times radius^2$
   - **Trapezoid**: Area= $\frac{1}{2} \times (base_1 + base_2) \times height$

3. **Add the Areas Together:** Sum up the areas of all the simple shapes to get the total area of the compound shape.

Example:

* Calculate the area of the shape in front:

Solution:

This shape is composed of a semicircle and a trapezoid, and we need to calculate the area of each separately and then add them together:

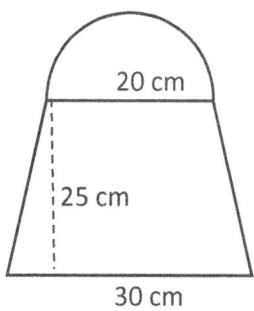

- Area of Semicircle: $\pi \times radius^2 \div 2 = \pi \times 10^2 \div 2 \approx 157 \ cm^2$
- Area of Trapezoid: $\frac{1}{2} \times (base_1 + base_2) \times height = \frac{1}{2} \times (20 + 30) \times 25 = 625 \ cm^2$
- Add the Areas Together: $area\ of\ semicircle + area\ of\ trapezoid = 157 + 625 = 782\ cm^2$.

www.mathnotion.com

Cubes

A cube is a three-dimensional geometric shape that consists of six equal square faces, twelve equal edges, and eight vertices. All angles in a cube are right angles (90 degrees).

Properties of a Cube

- **Faces:** A cube has six faces, and each face is square.
- **Edges:** A cube has twelve edges, and all edges are equal in length.
- **Vertices:** A cube has eight vertices (corner points).
- **Angles:** All internal angles between adjacent faces are right angles (90 degrees).

Formula for a Cube

- **Volume:** The volume V of a cube is given by:

$$V = side^3$$

Where "side" is the length of one of the cube's edges.

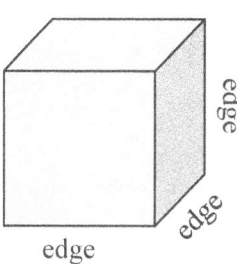

- **Surface Area:** The surface area A of a cube is given by:

$$A = 6 \times side^2$$

Examples:

1) A cube has an edge length of 5 inches. What is its volume and surface area?
   Solution:
   - The volume of a cube is given by:
     $V = side^3 = 5^3 = 5 \times 5 \times 5 = 125 in^3$.
   - The surface area of a cube is given by:
     $A = 6 \times side^2 = 6 \times 5^2 = 6 \times 5 \times 5 = 150 in^2$.

2) The surface area of the cube is $294 cm^2$. Find the length of cube.
   Solution:
   - The surface area of a cube is given by:
     $A = 6 \times side^2$ so, $294 = 6 \times side^2$
   - Solve for $side^2$:
     $$side^2 = 294 \div 6 = 49$$
   - Take the square root of both sides to find the edge length:
     $side = \sqrt{49} = 7$ cm.

www.mathnotion.com

Rectangular Prisms

A rectangular prism, also known as a cuboid, is a three-dimensional shape with six faces, all of which are rectangles. The shape is defined by its length, width, and height.

Characteristics of Rectangular Prisms:

- **Faces:** It has six rectangular faces.
- **Edges:** It has twelve edges.
- **Vertices:** It has eight vertices (corners).
- **Right Angles:** All interior angles between adjacent faces are right angles (90 degrees).

Properties of Rectangular Prisms:

Volume: The volume V of a rectangular prism can be calculated using the formula:

$$V = length \times width \times height$$

Surface Area: The surface area A of a rectangular prism is given by:

$$A = 2 \times (length \times width + width \times height + height \times length)$$

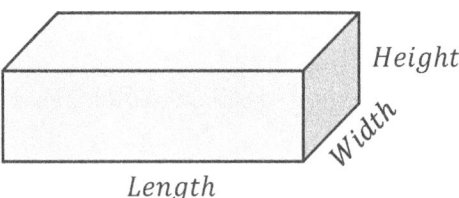

Example:

A rectangular prism has a length of 6 in, a width of 4 in, and a height of 5 in. Find its volume and surface area.

Solution:

1. Volume: The volume of a rectangular prism is given by:
   $V = length \times width \times height = 6 \times 4 \times 5 = 120$ cube inches.
2. The surface area A of a rectangular prism is given by:
   $A = 2 \times (length \times width + width \times height + height \times length)$
   $A = 2 \times (6 \times 4 + 4 \times 5 + 5 \times 6) = 2 \times (24 + 20 + 30) = 2 \times 74$
   $A = 148 \ in^2.$

Triangular Prisms

A triangular prism is a polyhedron with two parallel, congruent triangular bases connected by three rectangular faces. The shape is defined by the dimensions of its triangular base and the height (or length) between the bases.

Characteristics of Triangular Prisms:

1. **Bases:** Two congruent triangles that are parallel to each other.
2. **Faces:** Three rectangular faces that connect the corresponding sides of the triangular bases.
3. **Edges:** Nine edges in total – three for each triangular base and three connecting the corresponding vertices of the bases.
4. **Vertices:** Six vertices – three on each triangular base.

Properties of Triangular Prisms:

- **Volume:** The volume V of a triangular prism can be calculated using the formula:

$$V = Base\ Area \times Height$$

Where "Base Area" is the area of the triangular base and "Height" is the perpendicular distance between the two triangular bases.

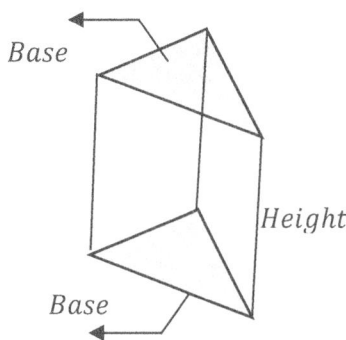

- **Surface Area:** The surface area A of a triangular prism is the sum of the areas of the two triangular bases and the three rectangular faces. It can be calculated by finding the area of each face and adding them together.

Example:

★ Find the volume of this triangular prism:

Solution:

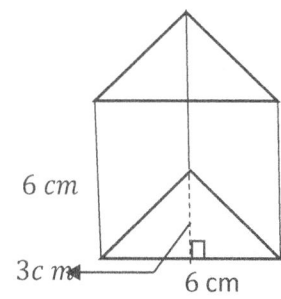

- Base Area: $\frac{1}{2} \times base \times height = \frac{1}{2} \times 6 \times 3 = 9$ square cm.
- Volume: $Base\ Area \times Height = 9 \times 6 = 54$ cubic cm.

Cylinder

A cylinder is a three-dimensional geometric shape with two parallel, congruent circular bases connected by a curved surface. It's a very common shape in geometry with various applications.

Characteristics of Cylinders:

- **Bases**: Two parallel, congruent circles.
- **Height (h):** The perpendicular distance between the two bases.
- **Radius (r):** The distance from the center to the edge of the circular base.
- **Curved Surface:** The surface that connects the two circular bases.

Properties of Cylinders:

- **Volume**: Volume V of a cylinder can be calculated using the formula:

$V = \pi r^2 h$ Where r is the radius of the base and h is the height of the cylinder.

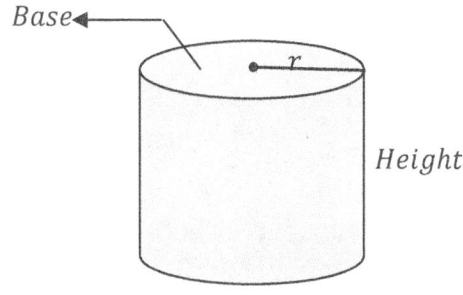

- **Surface Area:** The surface area A of a cylinder is the sum of the areas of its two bases and the curved surface. It can be calculated using the formula:

$$A = 2\pi r(r + h)$$

Where r is the radius of the base and h is the height of the cylinder.

Example:

A cylinder has a radius of 4 cm and a height of 10 cm. Find its volume and surface area.

Solution:

1. Volume: The volume of a cylinder is given by the formula:
$V = \pi r^2 h$
$V = \pi \times 4^2 \times 10 = \pi \times 16 \times 10 = 160\pi \approx 502.65$ cubic cm.

2. Surface Area: The surface area of a cylinder is given by the formula:
$A = 2\pi r(r + h)$
$A = 2 \times \pi \times 4 \times (4 + 10) = 8 \times \pi \times 14 = 112\pi \approx 351.68$ square cm.

Word Problems

Just like in other chapters, the process of solving word problems remains the same. However, in the context of geometry, we can add some extra steps. Here is a summary of this process

- **Read the Problem Carefully**: Understand what is being asked and the information provided.
- **Draw a Diagram:** Visualize the problem by drawing a diagram and labeling all known values and quantities to find.
- **Identify Relevant Formulas:** Recognize the geometric shapes involved and write down relevant formulas for area, perimeter, volume, and surface area.
- **Assign Variables:** Assign variables to unknown quantities and express the relationships between **known and unknown quantities using equations.**
- **Set Up Equations:** Use the information given in the problem to set up equations that are consistent with the identified formulas.
- **Solve the Equations**: Solve the equations step by step to find the values of unknown quantities. Check your calculations for accuracy.
- **Interpret the Solution:** Relate your solution back to the context of the problem to ensure it makes sense in the real-world context.
- **Verify the Solution**: Double-check your work by substituting the solution back into the original problem and ensuring it satisfies all conditions stated.

Example:

A rectangular garden has a length of 12 meters and a width of 8 meters. A path 1 meter wide is built around the garden. What is the total area of the garden including the path?

Solution:

1. Read the Problem Carefully: We need to find the total area of the garden including the path.
2. Draw a Diagram:
3. Identify Relevant Formulas: Area of rectangle= $length \times width$
4. Assign Variables: Garden length= 12 meters, garden width=8 meter and path width=1 meter
5. Set Up Equations:
   - New length including path: $12 + 2 \times 1 = 12 + 2 = 14 \, m$
   - New width including path: $8 + 2 \times 1 = 8 + 2 = 10 \, m$
6. Solve the equation: Total area including path: $14 \times 10 = 140 \, m^2$.
7. Interpret the Solution: The total area of the garden including the path is 140 square meters.
8. Verify the Solution: Double-check the calculations and ensure they align with the problem's context.

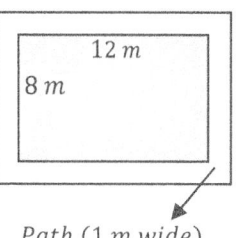

Path (1 m wide)

www.mathnotion.com

Worksheets

Introduction to Geometry

Lines, Segments, and Angles

✎ Identify whether each angle is an acute, obtuse, right or straight angle:

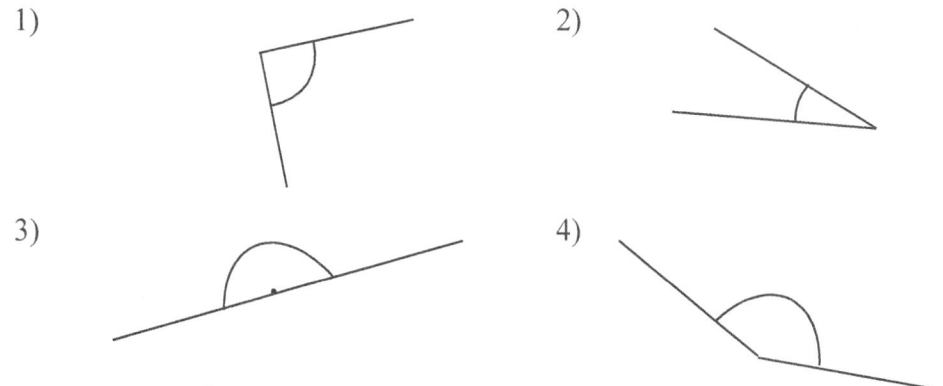

Complementary and Supplementary Angles

✎ Find the value of missing angle:

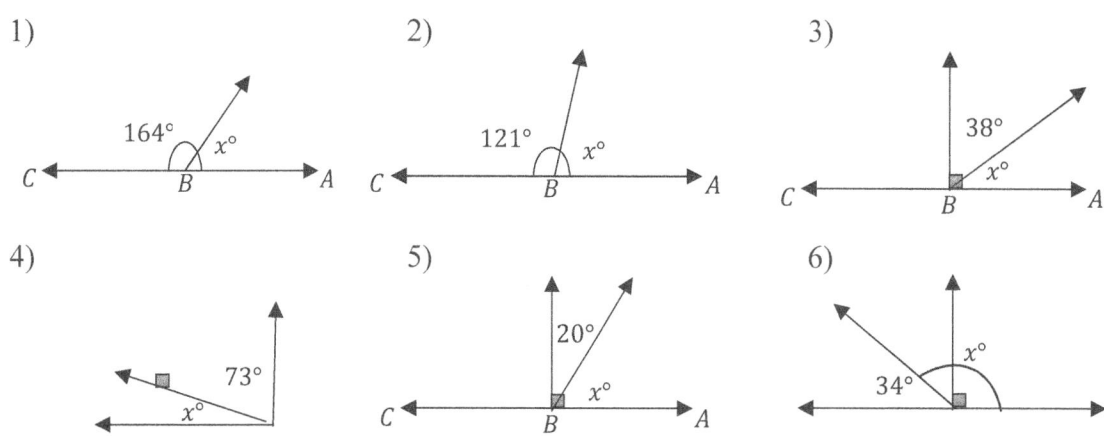

✎ Do the following problems:

7) Two supplement angles have equal measures. What is the measure of each angle?
8) The measure of an angle is seven fifth the measure of its supplement. What is the measure of the angle?
9) Two angles are complementary and the measure of one angle is 24 less than the other. What is the measure of the smaller angle?
10) Two angles are complementary. The measure of one angle is one fifth the measure of the other. What is the measure of the bigger angle?

Symmetry and Congruence

✏ Which type of symmetry do the following shapes have?

1)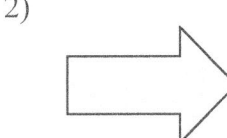
2)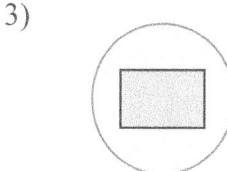
3)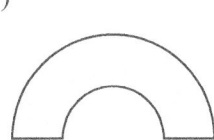

4)
5)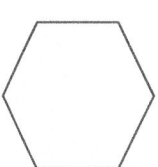
6)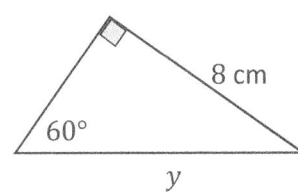

7) Circle A and Circle B are congruent. The radius of Circle A is 3 cm. What is the circumference and area of Circle B?

8) Rectangle PQRS is congruent to Rectangle TUVW. The dimensions of Rectangle PQRS are 7 cm by 5 cm. If side TU of Rectangle TUVW is 7 cm, what is the length of side TV?

9) The two following triangles are congruent. Find x, y and z.

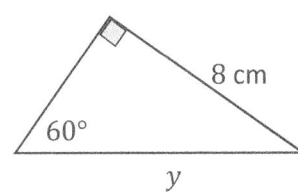 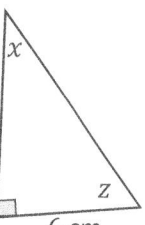

10) Two following polygons are congruent. Find x, y and z. (units are centimeter)

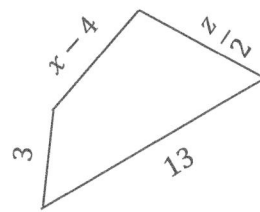 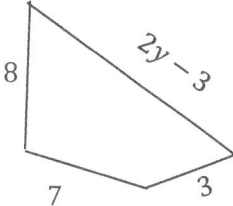

Classifying Polygons

✏ Identify the regularity and convexity of the following shapes:

1)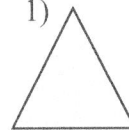
2)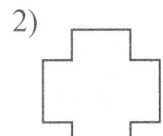
3)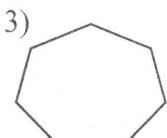
4)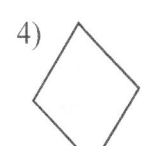
5)

Classifying Triangles

✎ Which kind of triangle are following triangles? (Equilateral, Isosceles, Scalene, Acute, Right or obtuse)

1) 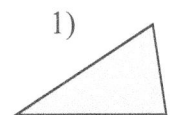   2) 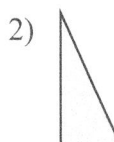   3) 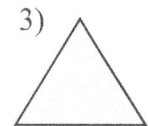   4) 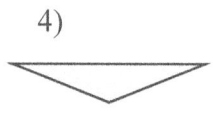   5)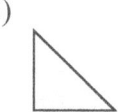

Pythagorean Relationship

✎ Find the missing values.

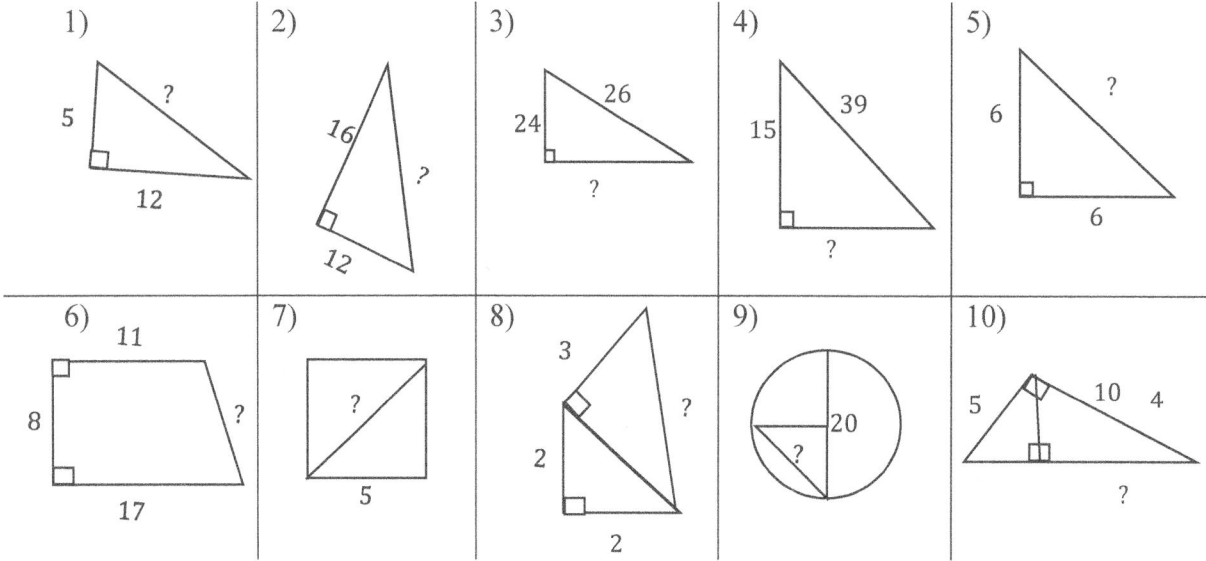

Area and Perimeter

Rectangles

✎ Find the area and perimeter of the following rectangles: (units are inch)

1) 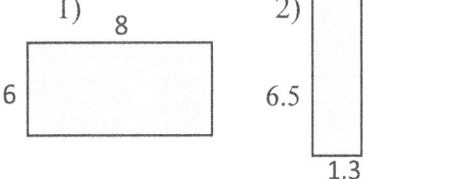   2) 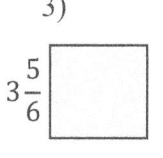   3) $3\frac{5}{6}$ 4) 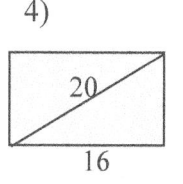   5)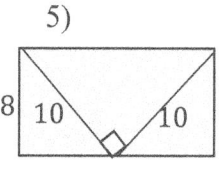

✎ Do the following problems with rectangles:

6) The area of the rectangle is 45 squares meters, and its width is 5 meters. Find its length and perimeter.

7) A rectangle has a perimeter of 36 cm and a length that is 4 cm longer than its width. Find the dimensions of the rectangle and its area.

8) A rectangle has an area of 60 in^2 and a diagonal of 13 in. Find its length and width.

9) A rectangle has a length that is 3 cm more than twice its width. If the perimeter of the rectangle is 54 cm, find its dimensions and the area of the rectangle.

www.mathnotion.com

10) Find a rectangle with an area of 48 square centimeters whose sides are integers and whose perimeter is the minimum possible.

Parallelograms

✎ Find the missing values about following parallelograms:

1)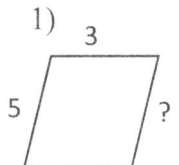
2)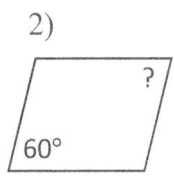
3)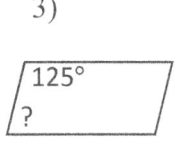
4)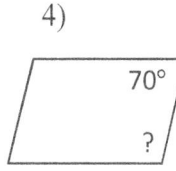
5)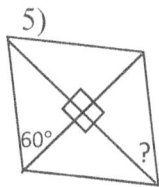

✎ Do the following problems about parallelograms:

6) A parallelogram has a base of 10 in and a height of 5 in. Find its area.
7) A parallelogram has sides of 6 cm and 8 cm. Find its perimeter.
8) A parallelogram has a base of 12 cm, a height of 7 cm, and one of its sides is 9 cm long. Find its area and perimeter.
9) A parallelogram has a base of 14 in and a height of 6 in. One of its sides (different from the base) is 10 in long. Find the area and the perimeter of the parallelogram.
10) The area of a parallelogram is 200 in^2, and its base is twice the length of its height. If the perimeter of the parallelogram is 56 in, find the dimensions of the parallelogram (base and height).

Triangles

✎ Find the missing value.

1)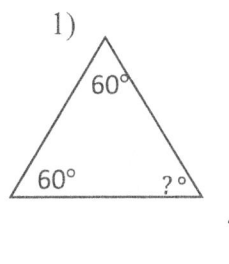
2)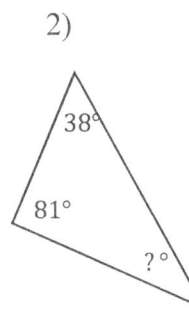
3)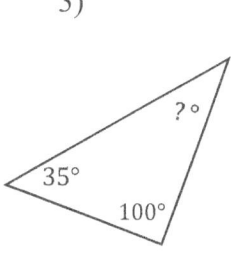
4)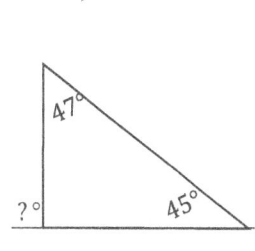
5)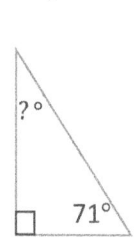

✎ Do the following problems about triangles:

6) A triangle has a base of 8 in and a height of 5 in. Find its area.
7) A triangle has sides of 3 cm, 4 cm, and 5 cm. Find its perimeter.
8) An equilateral triangle has a side length of 6 m. Find its perimeter.
9) A right triangle has legs of 9 in and 12 in. Find its area and the length of the hypotenuse.
10) An isosceles triangle has two equal sides of 10 cm each and a base of 12 cm. Find the area and the perimeter of the triangle.

Trapezoids

✎ Find the area of each trapezoid: (units are inch)

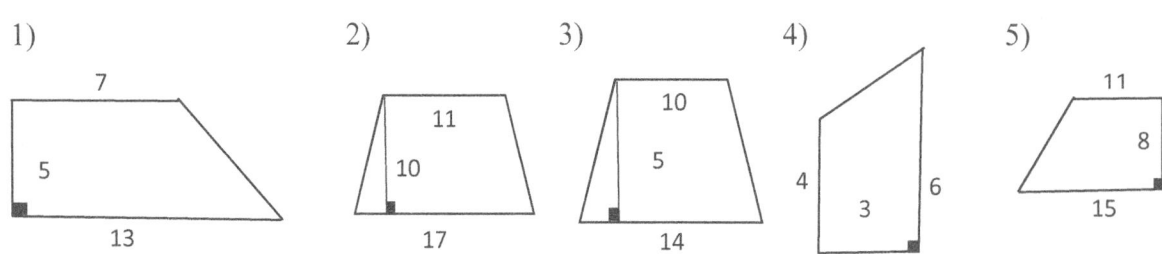

✎ Do the following problems about triangles:

6) A trapezoid has an area of 45 cm^2 and its height is 5 cm, and one base is 5 cm. What is the other base length?
7) If a trapezoid has an area of 99 ft^2 and the lengths of the bases are 8 ft and 10 ft, find the height?
8) If a trapezoid has an area of 126 m^2 and its height is 14 m and one base is 6 m, find the other base length?
9) The area of trapezoid is 440 ft^2 and its height is 22 ft. If one base of the trapezoid is 15 ft, what is the other base length?
10) The perimeter of a trapezoid is 22 cm, if its height is 4 cm and the sum of 2 legs is 6, calculate its area.

Circles

✎ Find the area and circumference of following circles: ($\pi \approx 3.14$ and units are foot)

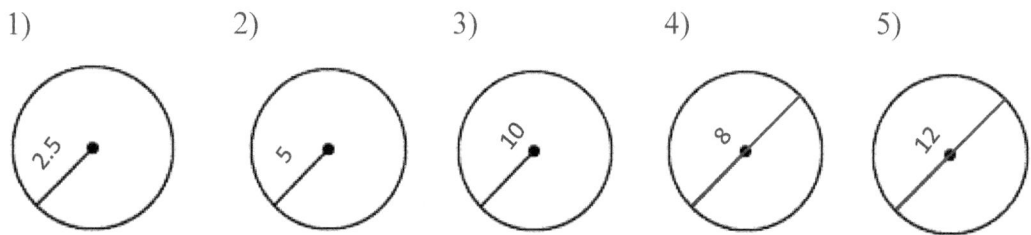

✎ Do the following problems about circles:

6) A circle has a circumference of 42 ft. Find its radius. ($\pi \approx 3$)
7) The area of a circle is 243 in^2. Find its radius. ($\pi \approx 3$)
8) A circle has an area of 75 ft^2. Find its circumference. ($\pi \approx 3$)
9) The circumference of a circle is 60 cm. Find its area. ($\pi \approx 3$)
10) A solid semicircle has a diameter of 10 in. Find its circumference. ($\pi \approx 3$)

Area of Compound Shapes

✎ Find the area of following compound shapes: ($\pi \approx 3.14$ and units are inch)

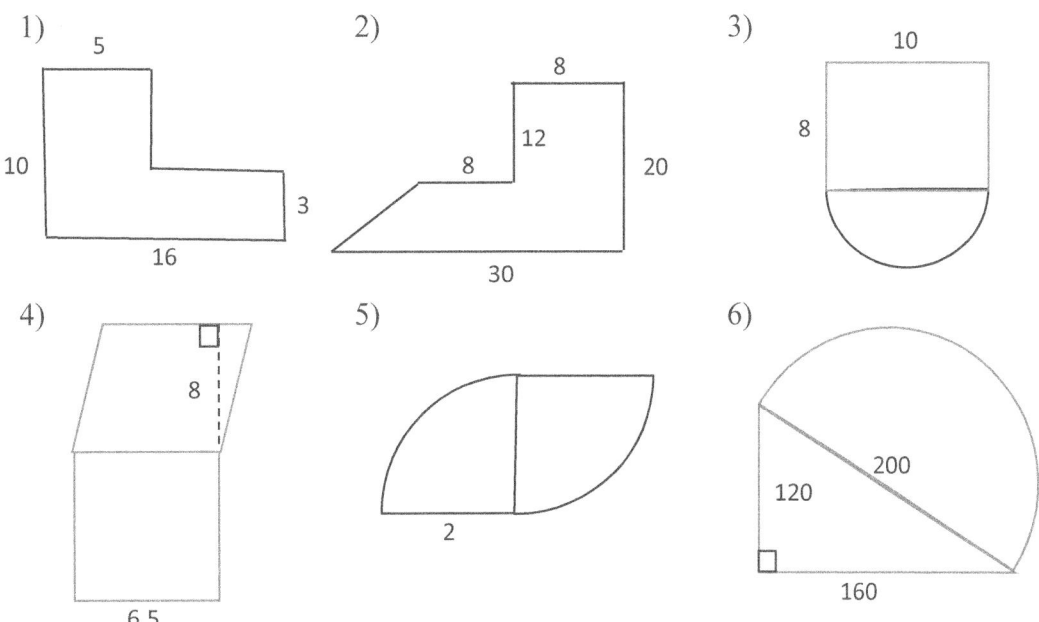

🖋 Find the area of gray part: ($\pi \approx 3.14$ and units are inch)

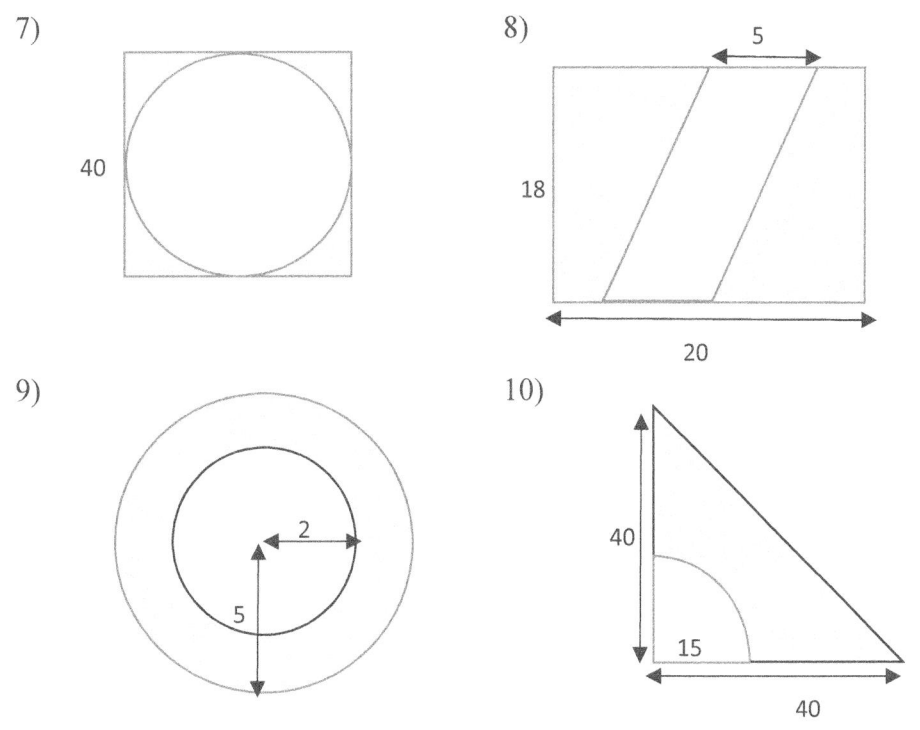

Surface Area and Volume

Cubes

🖋 Find the volume and surface area of the following cubes: (units are meters)

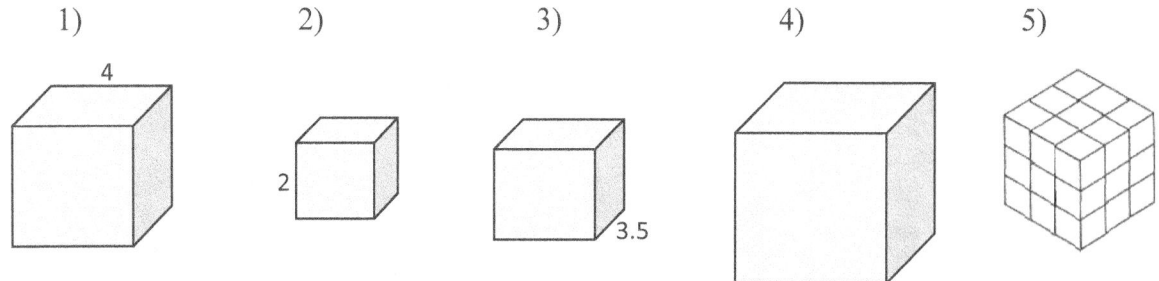

🔖 Do the following problems about cubes:

6) A cube has a volume of 64 cm^3. Find its side length.

7) A cube has a surface area of 54 in^2. Find its side length.

8) A cube has a volume of 512 cubic meters. Find its surface area.

9) The total length of all the edges of a cube is 60 ft. Find its volume.

10) A cube has a surface area that is 6 times its volume. Find the side length of the cube.

Rectangular Prisms

🔖 Find the volume and surface area. (units are foot)

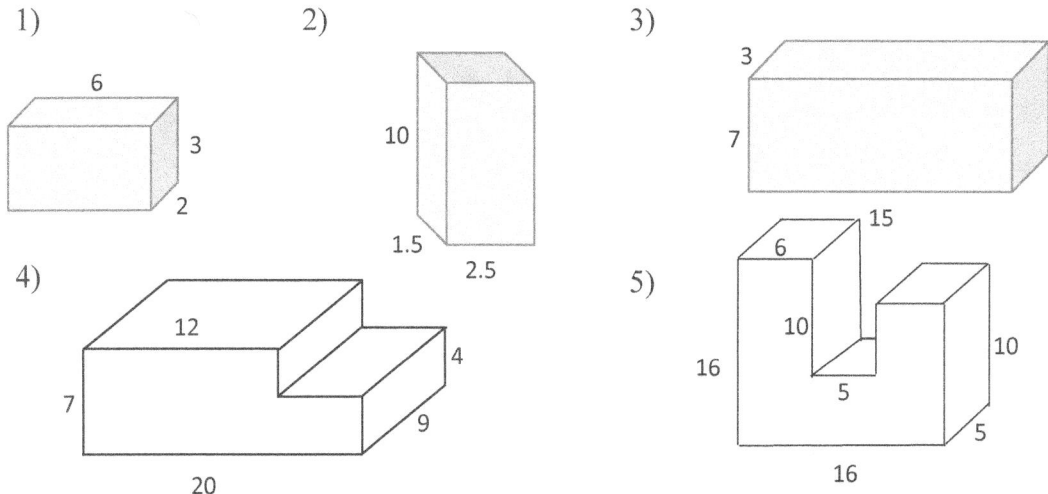

🔖 Solve.

6) A rectangular prism has a volume of 120 cm^3. If its length is 5 cm and its width is 4 cm, find its height.

7) A rectangular prism has a surface area of 94 ft^2. If its length is 7 ft and its width is 3 ft, find its height.

8) The length of a rectangular prism is 3 times its width, and its height is 4 cm. If the volume of the prism is 108 cubic centimeters, find its dimensions.

9) The total length of all the edges of a rectangular prism is 72 in. If its length is 10 in, and its width is twice its height, find the dimensions of the prism.

www.mathnotion.com

10) If we subtract 10% from the length, width, and height of a rectangular cube, how many percent will be reduced from its volume and area?

Triangular Prisms

✎ Find the volume and surface area of each triangular prism: (units are centimeter)

1)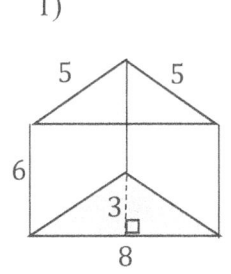
2)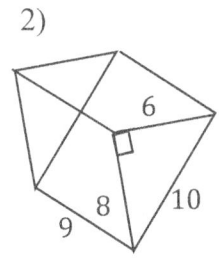
3)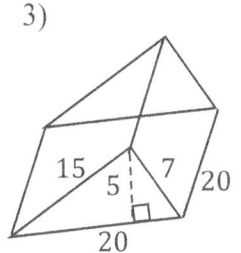
4)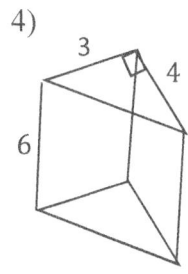
5)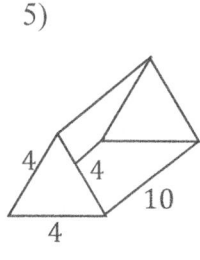

✎ Do following problems about triangular prism:

6) A triangular prism has a right-angled triangle as its base. The base has legs measuring 6 cm and 8 cm, and the height of the prism is 12 cm. What is the volume of the prism?
7) A triangular prism has a triangular base with sides 8 cm, 10 cm, and 12 cm, and the height of the prism is 14 cm. Find the lateral area of the triangular prism.
8) Calculate the height of a triangle prism if the area of its base is 16 squares meters and its volume is 48 cubic meters.
9) Calculate the perimeter of a triangular prism's base, if its height is 24 ft and its lateral area is 72 ft^2.
10) We have two triangular prisms, A and B. The area of the base of prism A is twice that of prism B, and the height of prism B is twice that of prism A. Which prism has a larger volume?

Cylinder

✎ Can you find the volume and surface area of each cylinder? ($\pi \approx 3.14$ and units are inch)

1)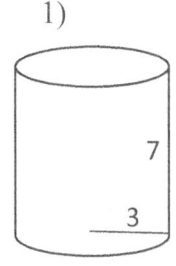
2)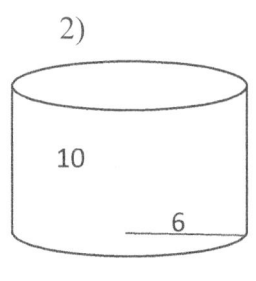
3)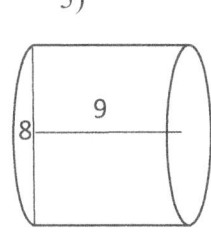
4)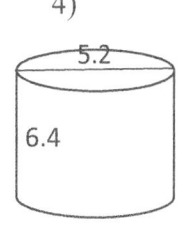
5)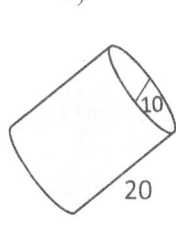

✎ Do the following problems about cylinder. ($\pi \approx 3$)

6) A cylinder has a radius of 7 cm and a height of 10 cm. Find its volume and surface area.
7) A cylinder has a volume of 1,200 cubic inches and a height of 4 in. Find its radius.
8) A cylinder has a volume of 1,500 ft^3 and a radius of 5ft. Find its height.

www.mathnotion.com

9) The surface area of a cylinder is 1,800 cm^2, and its height is twice its radius. Find its radius and height.

10) If we double the radius of a cylinder, by what factor will its volume increase?

Word problems

✍ Solve.

1) At 3:30 PM, the hands of a clock form an angle. What is the angle between the hour hand and the minute hand at 3:30?

2) A circular track has a radius of 10 meters. John wants to run around the track twice. How far will he run in total? ($\pi \approx 3$).

3) A ladder leans against a wall. The bottom of the ladder is 6 feet away from the wall, and the ladder reaches a height of 8 feet on the wall. How long is the ladder?

4) A shipping company needs to package small cubes with a side length of 3 cm. If each package can hold one cube, what is the volume of each cube?

5) A ramp is being built for a wheelchair. The ramp will have a rise of 3 feet and a horizontal run of 4 feet. How long will the ramp take from the ground to the door?

6) A fish tank is in the shape of a rectangular prism with a length of 30 cm, a width of 15 cm, and a height of 20 cm. How much water can the tank hold? ($\pi \approx 3$).

7) A storage box is in the shape of a rectangular prism with a volume of 2400 cubic cm. If the base area is 300 square cm, how tall is the box?

8) A can of soup has a radius of 3.5 cm and a height of 12 cm. How much metal is needed to make the can? ($\pi \approx 3$).

9) In a large box with dimensions 80 x 90 x 60 cm, what is the maximum number of tissue boxes with dimensions 15 x 10 x 6 cm that can fit?

10) If we want to pave around a large field with a diameter of 8 meters at 1.5 meters from the edge of the field, how many square meters of paving stones will we need? ($\pi \approx 3$).

Answer of Worksheets

Introduction to Geometry

Lines, Segments, and Angles

1) Right
2) Acute
3) Straight
4) obtuse

Complementary and Supplementary Angles

1) 16°
2) 59°
3) 52°
4) 17°
5) 70°
6) 146°
7) 90°
8) 105°
9) 33°
10) 75°

Symmetry and Congruence

1) Line (5 lines) and rotational (72°, 144°,…) symmetry.
2) Line (1line) symmetry.
3) Line (2 lines), rotational (180°) and point symmetry.
4) Line (1 line) symmetry.
5) Line (1 line) symmetry.
6) Line (6 lines), rotational (60°, 120°,…) and point symmetry.
7) Circumference ≈ 18.84 cm and area ≈ 28.26 square cm.
8) 5 cm.
9) $x = 30°, y = 10$ units and $z = 60°$.
10) $x = 11$ cm, $y = 8$ cm and $z = 16$ cm.

Classifying Polygons

1) Regular and convex.
2) Irregular and concave.
3) Regular and convex.
4) Irregular and convex.
5) Irregular and concave.

Classifying Triangles

3) Scalene, and acute.
4) Scalene and right.
5) Equilateral and acute.
6) Isosceles and obtuse.
7) Isosceles and right.

Pythagorean Relationship

1) 13 2) 20 3) 10 4) 36 5) $\sqrt{72}$
6) 10 7) $\sqrt{50}$ 8) $\sqrt{17}$ 9) $\sqrt{200}$ 10) $3+\sqrt{84}$

Rectangles

1) Area: 48 square in and Perimeter: 28 in.
2) Area: 8.45 square in and Perimeter: 15.6 in.
3) Area: $\frac{529}{36}$ square in and Perimeter: $15\frac{1}{3}$ in.
4) Area: 192 square in and Perimeter: 56 in.
5) Area: 96 square in and Perimeter: 40 in.
6) Length: 9 m and Perimeter: 28 m.
7) Length: 11 cm, $Width$: 7 cm and Area: 77 square cm.
8) Length: 12 in and $Width$: 5 in.
9) Length: 19 cm, $Width$: 8 cm and Area: 152 squares cm.
10) A rectangle with the dimensions of 6 cm by 8 cm.

Parallelograms

1) 5
2) 60°
3) 55°
4) 110°
5) 30°
6) 50 squares inches.
7) 28 cm.
8) Area = 84 square cm, Perimeter= 42 cm.
9) Area = 84 square inches, Perimeter= 48 inches.
10) Base=20 inches, Height=10 inches and the other side=8 inches.

Triangles

1) 60°
2) 61°
3) 45°
4) 92°
5) 19°
6) 20 squares inches.
7) 12 cm.
8) 18 m.
9) Area=54 square inches, Hypotenuse=15 inches
10) Area=48 square cm, Perimeter=32 cm

Trapezoids

1) 50 squares inches.
2) 140 squares inches.
3) 60 squares inches.
4) 15 squares inches.
5) 104 squares inches.
6) 13 cm.
7) 11 feet.
8) 12 meters.
9) 25 feet.
10) 32 square cm.

Circles

1) 19.625 square ft.
2) 78.5 square ft.
3) 314 squares ft.
4) 50.24 square ft.
5) 113.04 square ft.
6) 7 ft.
7) 9 inches.
8) 30 ft.
9) 300 squares cm.
10) 25 inches.

Area of Compound Shapes

1) 83 squares inches.
2) 280 squares inches.
3) 119.25 square inches.
4) 94.25 square inches.
5) 6.28 square inches.
6) 25,300 squares inches.
7) 344 squares inches.
8) 270 squares inches.
9) 65.94 square inches.
10) 623.375 square inches.

Cubes

1) Volume= 64 cubic m, Surface area=96 square m.
2) Volume= 8 cubic m, Surface area=24 square m.
3) Volume= 42.875 cubic m, Surface area=73.5 square m.
4) Volume= 216 cubic m, Surface area=216 square m.
5) Volume= 27 cubic m, Surface area=54 square m.
6) 4 cm.
7) 3 inches.
8) 384 squares m.
9) 125 cubic ft.
10) 1 unit.

Rectangular Prisms

1) Volume= 36 cubic ft, Surface area=72 square ft.
2) Volume= 37.5 cubic ft, Surface area=87.5 square ft.
3) Volume= 315 cubic ft, Surface area=342 square ft.
4) Volume= 1,044 cubic ft, Surface area=718 square ft.
5) Volume= 880 cubic ft, Surface area=712 square ft.
6) 6 cm.
7) 2.6 ft.
8) Length: 9 cm, Width: 3 cm, Height: 4 cm.
9) Length: 10 in, Width: 5.33 in, Height: 2.67 in.
10) 27.1% reduction in its volume and a 19% reduction in its surface area.

Triangular Prisms

1) Volume= 72 cubic cm, Surface area=132 square cm.
2) Volume= 216 cubic cm, Surface area=264 square cm.
3) Volume= 1,000 cubic cm, Surface area=940 square cm.

4) Volume= 36 cubic cm, Surface area=84 square cm.
5) Volume= $40\sqrt{3}$ cubic cm, Surface area=$8\sqrt{3}$ +120 square cm.
6) 288 cubic cm.
7) 420 squares cm.
8) 3 m.
9) 3 feet.
10) Equal volume.

Cylinder

1) Volume= 197.82 cubic in, Surface area=188.4 square in.
2) Volume= 1,130.4 cubic in, Surface area=602.88 square in.
3) Volume= 452.16 cubic in, Surface area=326.56 square in.
4) Volume= 135.84896 cubic in, Surface area=146.952 square in.
5) Volume= 1,570 cubic in, Surface area=785 square in.
6) Volume= 1,470 cubic cm, Surface area=714 square cm.
7) 10 inches.
8) 20 feet.
9) Radius=10 cm, Height=20 cm.
10) The new volume is 4 times the original volume.

Word problems

1) 75 degrees.
2) 120 meters.
3) 10 feet.
4) 27 cubic cm.
5) 5 feet.
6) 9,000 cubic cm.
7) 8 cm.
8) 325.5 square cm.
9) 480 boxes.
10) 42.75 square meters.

Chapter 13: Statistics and Probabilities

Topics that you will learn in this chapter:

- Mean and Median
- Mode and Range
- Histograms
- Line Graphs
- Times Series
- Stem-and-Leaf Plot
- Quartile of a Data Set
- Box-and- Whisker Plots
- Dot Plots
- Pie Graph
- Counting Principle
- Probability of Simple Events
- Probability of Opposite Events
- Making Predictions
- Word Problems
- Worksheet

Mean and Median

Mean: The Mean, also known as the average, is the sum of all the numbers in a dataset divided by the number of values in that dataset. It's a useful way to get a sense of the overall "middle" value of your data.

How to Calculate the Mean:

1. Add up all the numbers in the dataset.
2. Divide the sum by the number of values.

Median: The median is the middle value of a dataset when it is ordered from least to greatest. If there is an odd number of values, the median is the middle one. If there is an even number of values, the median is the average of the two middle numbers.

How to Calculate the Median:

1. Arrange the numbers in ascending order.
2. Find the middle value (or the average of the two middle values if the dataset has an even number of values).

Examples:

1) Find the Mean of following dataset: 1, 8, 5, 7, 11, 15
   Solution:
   - Add up all the numbers in the dataset: $4 + 8 + 5 + 7 + 9 + 15 = 48$
   - Divide the sum by the number of values: $48 \div 6 = 8$

   So, the Mean of this dataset is 8.

2) Calculate the median of following dataset: 9, 3, 8, 6, 7, 3, 1
   Solution:
   - Arrange the numbers in ascending order: 1, 3, 3, 6, 7, 8, 9
   - Find the middle value: The middle value is 6.

   So, the median of this dataset is 6.

3) Calculate the median of following dataset: 14, 5, 0, 8, 9, 6
   Solution:
   - Arrange the numbers in ascending order: 0, 5, 6, 8, 9, 14
   - The average of the two middle values: $(6 + 8) \div 2 = 7$

   So, the median of this dataset is 7.

Mode and Range

Mode: The mode is the value that appears most frequently in a dataset. A dataset may have one mode, more than one mode, or no mode at all if no number repeats.

How to Find the Mode:

1. List all the numbers in the dataset.
2. Identify the number(s) that appear most frequently.
- ☑ If two or more numbers appear with the same highest frequency, the dataset is multimodal (having multiple modes).

Range: The range is a measure of how spread out the values in a dataset are. It is the difference between the highest and lowest values.

How to Find the Range:

2. Identify the highest (maximum) value in the dataset.
3. Identify the lowest (minimum) value in the dataset.
4. Subtract the minimum value from the maximum value.

Examples:

1) Find the mode of following dataset: 5, 8, 12, 6, 1, 6, 1, 8, 1

   Solution:

   The mode is the number that appears the most:
   - 5 appears once.
   - 8 appears twice.
   - 12 appears once.
   - 6 appears twice.
   - 1 appears three times.

   Mode: 1 (since it appears three times, more than any other number).

2) Calculate the range of the same dataset: 5, 8, 12, 6, 1, 6, 1, 8, 1

   Solution:
   1. Maximum value: 12
   2. Minimum value: 1
   3. Range: $12 - 1 = 11$.

Histograms

Histograms are a type of graphical representation used in statistics and probability to visualize the distribution of a dataset. They help show the frequency of data points within specified ranges, known as bins.

Key Features of Histograms

1. **Bins (or Intervals):**
   - The horizontal axis (X-axis) represents the bins or intervals into which the data is divided.
   - Each bin covers a range of values, and the width of the bins is typically uniform.
2. **Frequency:**
   - The vertical axis (Y-axis) represents the frequency of data points within each bin.
   - The height of each bar indicates the number of data points that fall within that bin's range.
3. **Bars:**
   - Bars are used to depict the frequency of each bin.
   - The height of the bar corresponds to the count of data points within that bin's range.

How to Create a Histogram

Collect Data: Gather the dataset you want to analyze.

Determine Bins: Decide on the number of bins and their ranges. The choice of bins can affect the shape and interpretation of the histogram.

Count Frequencies: Count how many data points fall into each bin.

Draw the Bars:
- Plot the bins on the X-axis and the corresponding frequencies on the Y-axis.
- Draw bars for each bin with heights representing the frequencies.

Example:

Create a histogram for following data on exam scores: 55, 60, 65, 65, 70, 75, 80, 85, 90, 95

Solution: we can create a histogram with 5 bins and a range of 8 units. Then count data points fall into each bin:

- Frequency of data between 55 and 63: 2 (55, 60)
- Frequency of data between 63 and 71: 3 (65, 65, 70)
- Frequency of data between 71 and 79: 1 (75)
- Frequency of data between 79 and 87: 2 (80, 85)
- Frequency of data between 87 and 95: 2 (90, 95)

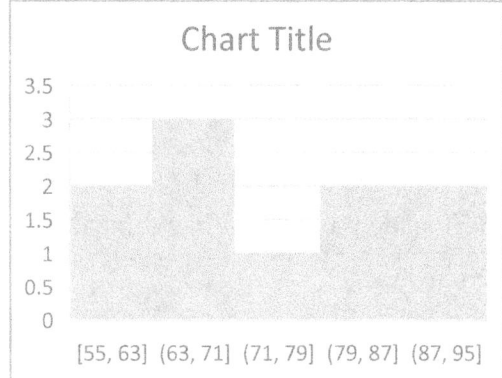

Line Graphs

Line graphs are a type of chart used to display information that changes over time or to show the relationship between two variables. They are particularly useful in statistics for visualizing trends and patterns.

Key Features of Line Graphs

1. **Axes:**
   - The horizontal axis (X-axis) typically represents time or the independent variable.
   - The vertical axis (Y-axis) represents the dependent variable.
2. **Points:** Data points are plotted on the graph based on the values of the variables.
3. **Lines:** Points are connected by straight lines to show how the values change.

How to Create a Line Graph

1. **Collect Data:** Gather the data that you want to plot
2. **Draw Axes:**
   - Draw the X-axis and Y-axis on graph paper or using graphing software.
   - Label the axes with the variables you're using.
3. **Plot Points:** For each pair of values, place a dot at the corresponding position on the graph.
4. **Connect Points:** Use a ruler to draw lines connecting the points in the order of the data.

Example:

If you have data on the number of books read over five months:

Month	January	February	March	April	May
Book Read	3	4	5	2	6

Draw a line graph based on the information above:

Solution:

Draw and Label Axes: X-axis: Months (January, February, March, April, May) and Y-axis: Number of Books Read

Plot Points: (January 3), (February 4), (March 5), (April 2), (May 6)

Connect Points: Draw lines connecting these points in order from January to May.

Times Series

Time series are a way of recording data points over a period of time, usually at regular intervals. It's a useful concept in statistics to track changes and identify patterns or trends over time.

Key Features of Time Series
1. **Time Intervals:** Data is collected at consistent time intervals, such as daily, weekly, monthly, or yearly.
2. **Data Points:** Each data point represents the value of the variable being measured at a specific time.
3. **Visualization:** Time series are often visualized using line graphs to show how the values change over time.

How to Create a Time Series Graph
1. **Collect Data:** Gather the data that you want to plot
2. **Draw Axes:**
   - Draw the X-axis and Y-axis on graph paper or using graphing software.
   - Label the axes with the variables you're using.
3. **Plot Points:** For each pair of values, place a dot at the corresponding position on the graph.
4. **Connect Points:** Use a ruler to draw lines connecting the points in the order of the data.

Example

Imagine you are tracking the temperature in your city every day for a week. Your data might look like this:

Day	Monday	Tuesday	Wednesday	Thursday	Friday	Saturday	Sunday
Temperature	15	17	16	18	14	19	17

Draw a Time series graph based on information above:

Solution:

1. Draw Axes:
   - Draw the horizontal axis (X-axis) to represent the days of the week.
   - Draw the vertical axis (Y-axis) to represent the temperature in degrees Celsius.
2. Plot Points: For each day, place a point at the corresponding temperature.
3. Connect Points: Draw lines to connect the points in order, creating a line graph.

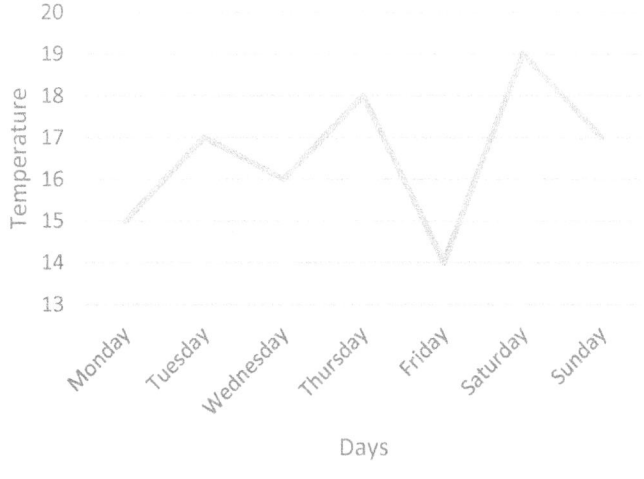

Stem-and-Leaf Plot

A stem-and-leaf plot is a way to organize and display data so it's easier to see the distribution and frequency of the values.

Key Features of Stem-and-Leaf Plots

1. **Stems.**
   - The "stem" represents the leading digits of the numbers.
   - In a dataset where the numbers range from 10 to 99, the tens place could serve as the stem.
2. **Leaves:**
   - The "leaf" represents the trailing digit of the numbers.
   - In the same dataset, the ones place would be the leaf.

How to Create a Stem-and-Leaf Plot

1. **Organize Data:** Collect your data and list it in ascending order.
2. **Determine Stems:** Identify the stems based on the leading digits of your data.
3. **List Leaves:** For each stem, list the trailing digits (leaves) in ascending order next to the corresponding stem.

Example:

✵ Create a Stem-and-Leaf plot for following dataset of test scores:

$$81, 35, 46, 31, 66, 83, 76, 88, 63, 69, 33, 78, 44$$

Solution:

1. Organize Data: Collect data in ascending order:
   31, 33, 35, 44, 46, 63, 66, 69, 76, 78, 81, 83, 88
2. Determine Stems: Stems are the tens place digits: 3, 4, 6, 7, 8
3. List Leaves: For each stem, list the ones place digits:

Stem	Leaf
3	1, 3, 5
4	4, 6
6	3, 6, 9
7	6, 8
8	1, 3, 8

Quartile of a Data Set

Quartiles are values that divide a data set into four equal parts, each containing 25% of the data. They are useful in statistics for understanding the distribution of data.

Key Quartiles

1. **First Quartile (Q1):**
   - Also known as the lower quartile.
   - It separates the lowest 25% of the data from the rest.
   - Located at the 25th percentile of the data set.
2. **Second Quartile (Q2):**
   - Also known as the Median.
   - It separates the data into two equal halves, with 50% of the data below and 50% above.
   - Located at the 50th percentile of the data set.
3. **Third Quartile (Q3):**
   - Also known as the upper quartile.
   - It separates the highest 25% of the data from the rest.
   - Located at the 75th percentile of the data set.

How to Calculate Quartiles:

1. **Organize the Data:** Arrange the data in ascending order.
2. **Find the Median (Q2):** If the data set has an odd number of values, the median is the middle value and if the data set has an even number of values, the median is the average of the two middle values.
3. **Calculate Q1 and Q3:**
   - Q1 is the median of the lower half of the data (excluding the overall median if the number of data points is odd).
   - Q3 is the median of the upper half of the data (excluding the overall median if the number of data points is odd).

Example:

Calculate the quartiles of following data: 3, 7, 8, 12, 14, 18, 21, 22, 24, 26.
   Solution:
   1. Organize the Data: Already in ascending order: 3, 7, 8, 12, 14, 18, 21, 22, 24, 26
   2. Find the Median (Q2): Since there are 10 data points (an even number), the median is the average of the 5th and 6th values. $Median\ (Q2) = \frac{14+18}{2} = \frac{32}{2} = 16$
   3. Calculate Q1 and Q3: Lower half of the data: 3, 7, 8, 12, 14 and upper half of the data: 18, 21, 22, 24, 26
      - Q1 is the median of the lower half (3, 7, 8, 12, 14): $Q1 = 8$
      - Q3 is the median of the upper half (18, 21, 22, 24, 26): $Q2 = 22$

Box- and -Whisker Plots

Box-and-Whisker Plots:

A Box-and-Whisker Plot (often just called a Box Plot) summarizes a dataset by displaying its quartiles and the minimum and maximum values. It's particularly useful for showing the spread and skewness of the data.

Key Components:

1. **Minimum:** The smallest value in the dataset.
2. **First Quartile (Q1):** The median of the lower half of the data.
3. **Median (Q2):** The middle value of the data.
4. **Third Quartile (Q3):** The median of the upper half of the data.
5. **Maximum:** The largest value in the dataset.
6. **Interquartile Range (IQR):** The range between Q1 and Q3 (Q3 - Q1).
7. **Whiskers:** The distance from the minimum to Q1 and from Q3 to the maximum.

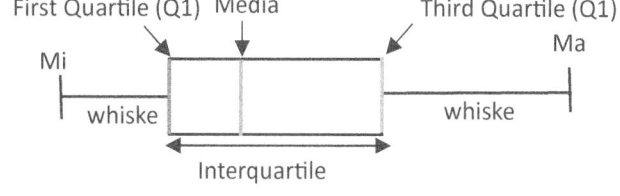

Example:

We have a dataset of the weights (in grams) of various apples, draw the box-and whisker plot for them. 120,135,140,145,150,160,165,170,175,185

Solution: First arrange the data in ascending order (it's already sorted) and then to draw the box-and-whisker plot we need all the following information:

- **Minimum:** 120
- **Median (Q2):** The middle value of the data: $\frac{150+160}{2} = \frac{310}{2} = 155$
- **First Quartile (Q1):** The median of (120, 135, 140, 145, 150) is 140
- **Third Quartile (Q3):** The median of (160, 165, 170, 175, 185) is 170
- **Maximum:** 185

Now draw a horizontal line that includes the range of your data from 110 to 190 and plot min, median, Q1, Q3 and max and then draw a box from Q1 (140) to Q3 (170). Finally, draw whiskers from the minimum (120) to Q1(140) and from Q3 (170) to the maximum (185):

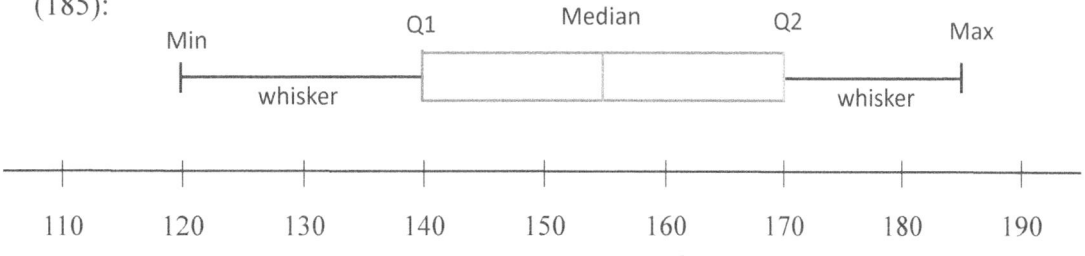

www.mathnotion.com

Dot Plots

A Dot Plot is a simple way to display data points on a number line. Each dot represents one data point, making it easy to see the frequency and distribution of the data.

Key Features:

- **Number Line:** A horizontal line representing the range of values.
- **Dots:** Each dot represents one occurrence of a value.

Example:

If the following data showing the number of books read by 17 students in a week, draw the dot plot about this data:

1, 1, 1, 1, 1, 1, 2, 3, 3, 4, 5, 5, 5, 6, 7, 8, 8

Solution:

1. Organize the Data: First, arrange the data in ascending order (it's already sorted).
2. Draw a Number Line: Draw a horizontal line and mark it with the values from the lowest to the highest number in the data set.
3. Plot the Data Points: Place a dot above each number for each time it occurs in the data set. For example, for the value 2, place one dot above the 2 on the number line or for the value 5, place three dots above the 5 on the number line:

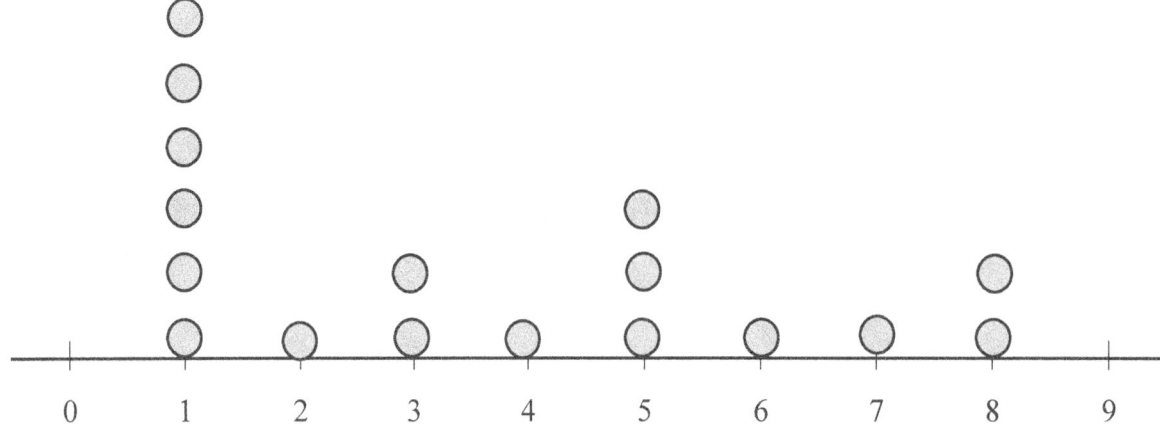

Pie Graph

A pie graph, also known as a pie chart, is a circular graph used to represent data. It's particularly useful for showing the proportions of different categories in a dataset.

Key Features of Pie Graphs

1. **Circle:** The entire circle represents the whole dataset (100%).
2. **Slices:**
   - The circle is divided into slices, each representing a different category.
   - The size of each slice is proportional to the percentage or fraction of the total that the category represents.
3. **Labels**: Each slice is usually labeled with the category it represents and its percentage or value.

How to Create a Pie Graph

1. **Collect Data:** Gather the data you want to represent.
2. **Calculate Percentages:** Determine what percentage of the total each category represents.
3. **Draw the Circle:** Draw a circle and divide it into slices based on the percentages calculated.
4. **Label the Slices:** Label each slice with the category and the corresponding percentage.

Example:

We have data on how a student spends 10 hours of their day on various activities:

activity	Studying	Exercising	Sleeping	Leisure (e.g., watching TV, playing games)
hour	4	1	3	2

Create a pie graph for these data.

Solution:

1. Calculate percentages:
   - Studying: $\frac{4}{10} = \frac{40}{100} = 40\%$
   - Exercising: $\frac{1}{10} = \frac{10}{100} = 10\%$
   - Sleeping: $\frac{3}{10} = \frac{30}{100} = 30\%$
   - Leisure: $\frac{2}{10} = \frac{20}{100} = 20\%$
2. Draw the Circle: Divide the circle into slices based on these percentages.
3. Label the Slices: Label each slice with the activity and its percentage.

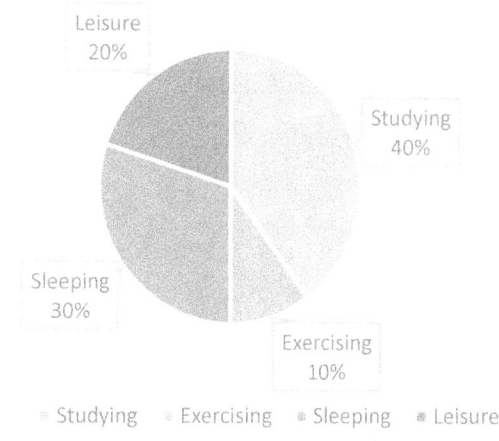

Counting principle

The Counting Principle helps us figure out how many ways we can combine choices when we have multiple options. It's like figuring out how many different outfits you can make if you have different shirts and pants to choose from.

How to Use the Counting Principle

- **Identify the Choices**: Determine the number of choices for each event.
- **Multiply the Choices:** Multiply the number of choices for each event together to get the total number of possible outcomes.

Examples:

1) If you have 2 choices for shirts (red, blue) and 3 choices for pants (jeans, shorts, black pants), how many different outfits can you make?

Solution:

1. List the Choices:
   - You have 2 choices of shirts (red, blue).
   - You have 3 choices of pants (jeans, shorts, black pants).
2. Multiply the Choices: Multiply the number of shirts by the number of pants:

   $2 \; shirts \times 3 \; pants = 6 \; outfits$

   This means you have 6 different combinations of shirts and pants to wear.

2) How many different two-digit numbers can we create by digits 1, 2, 3, and 4. Each digit can be used more than once.

Solution:

1. Identify the Choices for Each Digit:

   The first digit (tens place) can be any of the 4 digits: 1, 2, 3, 4.

   The second digit (one's place) can also be any of the 4 digits: 1, 2, 3, 4.
2. Multiply the Choices: Multiply the number of choices for the first digit by the number of choices for the second digit:

   $$4 \; (choices \; for \; tens \; place) \times 4 \; (choices \; for \; ones \; place)$$
   $$= 16 \; (different \; numbers)$$

   This means you can create 16 different two-digit numbers using the digits 1, 2, 3, and.

www.mathnotion.com

Probability of Simple Events

Probability of simple events is a basic concept in probability that helps us understand how likely it is for a specific event to happen.

Key Concepts

1. **Event:** An event is something that can happen. For example, rolling a die and getting a 4.
2. **Outcome:** An outcome is a possible result of an event. For example, the outcomes of rolling a die are 1, 2, 3, 4, 5, and 6.
3. **Probability:** Probability is a measure of how likely an event is to happen. It ranges from 0 to 1, where 0 means the event cannot happen, and 1 means the event is certain to happen.

How to Calculate Probability

The probability of a simple event is calculated using the formula:

$$\text{Probability} = \frac{\text{Number of favorable outcomes}}{\text{Total number of possible outcomes}}$$

Examples:

1) If we have a regular six-sided dice, find the probability of rolling a 4.
   Solution:
   1. Identify the Event: Rolling a 4
   2. Count the Favorable Outcomes: There is only 1 favorable outcome (rolling a 4).
   3. Count the Total Number of Possible Outcomes: There are 6 possible outcomes (rolling a 1, 2, 3, 4, 5, or 6)
   4. Calculate the Probability:

   $$\text{Probability} = \frac{1(\text{favorable outcomes})}{6\ (\text{possible outcomes})} = \frac{1}{6}$$

2) We have a bag with 5 red marbles and 3 blue marbles. Find the probability of randomly picking a blue marble.
   Solution:
   1. Identify the Event: Picking a blue marble.
   2. Count the Favorable Outcomes: There are 3 favorable outcomes (3 blue marbles).
   3. Count the Total Number of Possible Outcomes: There are 8 possible outcomes (5 red marbles + 3 blue marbles).
   4. Calculate the Probability:

   $$\text{Probability} = \frac{3(\text{favorable outcomes})}{8\ (\text{possible outcomes})} = \frac{3}{8}$$

www.mathnotion.com

Probability of opposite events

The probability of opposite events, also known as complementary events, are pairs of events where one event happens means the other cannot happen. In other words, the sum of the probability of these events is always 1 (or 100%).

Key Concept

If the probability of an event A happening is $P(A)$, then the probability of the event not happening (let's call it A' or the complement of A) is $P(A')$.

The formula is: $P(A') = 1 - P(A)$

Examples:

1) If you're tossing a fair coin, there are two possible outcomes: heads or tails. find the probability of **not** getting heads.
   Solution:
   1. Probability of Getting Heads (Event H): Since a coin has two sides, the probability of getting heads is: $P(H) = \frac{1}{2}$
   2. Probability of Not Getting Heads (Event H'): The complementary event is getting tails. Using the formula for complementary events:
   $$P(H') = 1 - P(H) = 1 - \frac{1}{2} = \frac{1}{2}$$

2) We have a standard deck of 52 playing cards, find the probability of **not** drawing an Ace.
   Solution:
   1. Probability of Drawing an Ace (Event A): There are 4 Aces in a deck of 52 cards, so the probability of drawing an Ace is: $P(A) = \frac{4}{52} = \frac{1}{13}$
   2. Probability of Not Drawing an Ace (Event A'): The complementary event is drawing any card that is not an Ace. Using the formula for complementary events:
   $$P(A') = 1 - P(A) = 1 - \frac{1}{13} = \frac{12}{13}$$

Word Problems

Steps to Solve Word Problems about statistics and probability:

1. **Read the Problem Carefully:** Understand what the problem is asking and identify key information and data provided in the problem.
2. **Define the Variables**: Determine what variables or quantities you need to find.
3. **Organize the Information:** List out the information given and what you need to find. Sometimes, drawing a diagram or a chart can help visualize the problem.
4. **Apply Relevant Formulas:**
   - For probability, this might include the formula for probability, complementary events, or combinations.
   - For statistics, this might include mean, median, mode, range, or other relevant measures.
5. **Solve the Problem Step by Step:** Break the problem down into smaller steps and solve each part of the problem logically and systematically.
6. **Check Your Answer:** Verify your calculations and ensure your answer makes sense in the context of the problem.

Example:

A bag contains 5 red marbles, 3 blue marbles, and 2 green marbles. What is the probability of randomly drawing a blue marble?

Solution:

1. Read the Problem: We need to find the probability of drawing blue marble.
2. Define the Variables:
   - Number of blue marbles = 3
   - Total number of marbles = 5 (red) + 3 (blue) + 2 (green) = 10
3. Organize the Information:
   - Number of favorable outcomes (blue marbles) = 3
   - Total possible outcomes (total marbles) = 10
4. Apply the Formula and Solve:

$$\text{Probability} = \frac{\text{Number of favorable outcomes}}{\text{Total number of possible outcomes}} = \frac{3}{10}$$

5. Check Your Answer: The calculation is correct, and the answer makes sense.

Worksheets

Mean and Median

🔖 Find the Mean.

1) 8, 6, 15, 9, 1, 1
2) 19, 25, 13, 31, 45, 11
3) 140, 141, 142, 143, 144, 145, 146
4) 5.5, 4.8, 1.2, 2.5
5) $\frac{1}{6}, 1\frac{2}{6}, 2, \frac{5}{6}, \frac{4}{6}, 1\frac{3}{6}, \frac{1}{2}$

🔖 Find the median.

1) 15, 6, 82, 71, 9, 21, 63, 7, 19, 33, 2
2) 25, 31, 52, 48, 16, 4
3) 12, 87, 6, 100, 5, 66
4) 7.5, 5, 3.8, 10, 1.7, 6.5, 3, 4.3
5) $\frac{5}{7}, 2, \frac{9}{2}, \frac{3}{2}$

Mode and Range

🔖 Find the mode of the following datasets:

1) 8, 2, 3, 1, 6, 4, 8, 5, 3, 3
2) 52, 14, 18, 14, 16, 14, 16, 18, 18, 14
3) 2, 2, 3, 1, 1, 2, 3, 3, 1, 1, 2, 3, 1, 3, 1
4) 34, 33, 30, 31, 30, 36, 38, 39, 30, 32, 33
5) 1.2, 3.5, 1, 4.5, 3.5, 8.5, 1, 2.5, 3.5, 3

🔖 Find the range of the following datasets:

1) 5, 7, 10, 125, 7, 10, 12
2) 15, 22, 8, 17, 10
3) 33, 28, 46, 50, 31, 29, 40
4) 18, 5, 12, 26, 7, 14, 19
5) 102, 88, 97, 110, 95, 85, 99

Histograms

🔖 Draw a histogram for the following datasets:

1) 5, 7, 10, 11, 15, 21 (group them into 4 bins)
2) 1, 2, 2, 3, 3, 3, 4, 4, 4, 4 (group them into 3 bins)
3) 18, 25, 30, 35, 40, 45, 50, 55, 60 (group them into 6 bins)

🔖 According to the following histogram, answer the questions 4-5:

4) Which bin has the lowest and highest frequency of data?
5) How much data is there in this dataset?

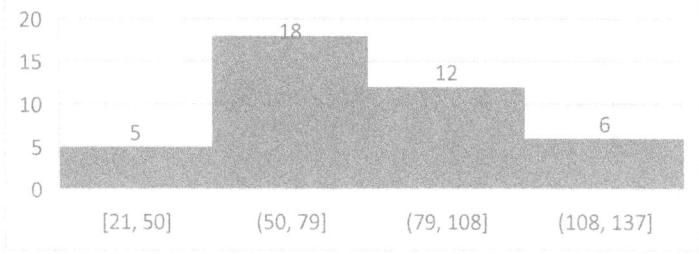

Times Series

✍ Draw a line graph based on the information in the tables:

1)

day	Monday	Tuesday	Wednesday	Thursday	Friday	Saturday	Sunday
temperature	15°	17°	16°	18°	19°	20°	21°

2)

week	Week 1	Week 2	Week 3	Week 4	Week 5	Week 6	Week 7	Week 8
Sales of a small shop in dollars	$100	$120	$140	$90	$160	$110	$180	$200

3)

month	January	February	March	Apri	May	June	July	August	September	October	November	December
Rainfall (in mm) in a city	50	40	60	55	70	80	75	90	85	65	50	45

✍ According to the following line graph, answer the questions 4-5:

4) Between which weeks did the plant have the most and least growth?
5) How much did the plant grow between the fifth and tenth weeks?

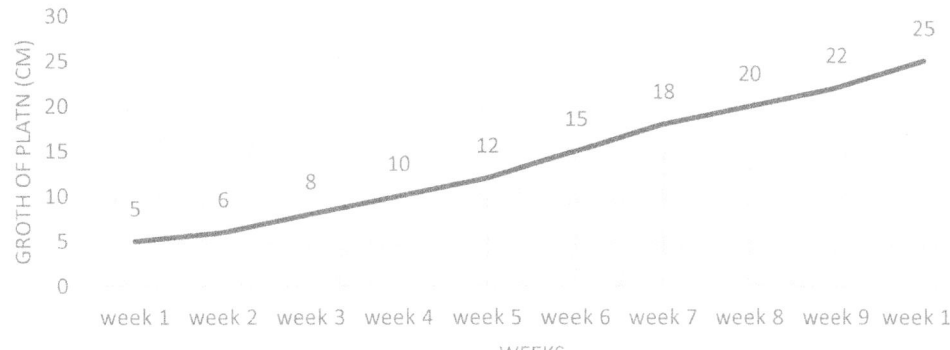

Stem-and-Leaf Plot

✍ Create a Stem-and-Leaf plot for the following datasets:

1) 12, 13, 18, 24, 25, 29, 30, 31
2) 43, 45, 47, 52, 56, 59, 60, 60, 63
3) 93, 95, 95, 102, 104, 106, 113, 114, 118, 119, 119
4) 120, 126, 131, 131, 139, 139, 142, 145
5) 254, 256, 259, 262, 265, 268, 273, 276, 279

www.mathnotion.com

Quartile of a Data Set

✎ Calculate the quartiles of the following datasets:

1) 4, 8, 15, 16, 23, 42
2) 12, 18, 22, 26, 31, 35, 42
3) 5, 9, 14, 20, 27, 33, 38, 42, 50
4) 8, 11, 13, 15, 17, 22, 24, 26, 29, 33, 35
5) 16, 21, 25, 28, 32, 34, 38, 41, 43, 47, 50, 52

Box-and- Whisker Plots

✎ Draw the box-and whisker plot for following dataset:

1) 5, 7, 8, 9, 10, 12, 14
2) 15, 18, 22, 25, 29, 31, 33, 38, 40
3) 28, 31, 35, 37, 40, 42, 45, 48, 50, 52
4) 55, 60, 65, 68, 72, 75, 78, 80, 82, 85, 89, 90
5) 102, 110, 115, 118, 122, 126, 130, 135, 138, 140, 145, 150, 155, 160

Dot Plots

✎ Draw the dot plot for following dataset:

1) 4,5,4,6,7,4,7
2) 2,3,5,2,4,3,5,2
3) 1,2,3,4,2,5,3,3,4,1
4) 10,12,11,14,15,12,13,10,13
5) 40, 45, 50, 45, 60, 55, 50, 40, 45, 50, 60, 55, 50, 45, 30

Pie Graph

✎ According to the following pie chart, answer the questions 1-2:

Monthly expenses of a family (dollars)

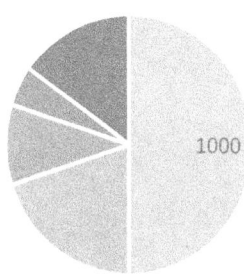

■ Rent ■ Groceries ■ Utilities ■ Transportation ■ Entertainment

1) How much is the total expenses of this family every month?
2) If families spend monthly 100 dollars on transportation, what percentage of the total expenses is related to transportation?

✎ Draw a pie chart based on the information in the tables:

3) Favorite fruit of a group of students:

Apples	Bananas	Oranges	Grapes
10	5	3	2

www.mathnotion.com

4) Number of pets owned by students in a class:

Dogs	Cats	Fish	Birds
8	6	4	2

5) Time spent on different activities in a day (in hours):

Sleeping	Studying	Playing	Watching	Other
8	4	2	2	8

Counting Principle

Do the following problems about counting principle:

1) How many different two-digit numbers can you create using digits 1, 2, and 3?
2) How many different ices cream sundaes can you make with 4 types of ice cream and 3 types of toppings?
3) A password consists of 2 letters followed by 2 digits. How many different passwords can be created using the letters A and B, and the digits 1 and 2?
4) How many different pizzas can be made if you can choose from 2 types of crusts, 3 types of sauces, and 4 types of toppings?
5) In a contest, you can choose from 3 appetizers, 4 main courses, and 2 desserts. How many different three-course meals can you create?

Probability of Simple Events

Do the following problems about counting principle:

1) What is the probability of rolling a 3 on a six-sided die?
2) What is the probability of drawing red marble from a bag containing 4 red marbles and 6 blue marbles?
3) What is the probability of drawing a queen from a standard deck of 52 playing cards?
4) A box contains 5 red, 3 green, and 2 blue balls. What is the probability of randomly selecting a green ball from the box?
5) A deck of cards has 52 cards, consisting of 13 hearts, 13 diamonds, 13 clubs, and 13 spades. What is the probability of drawing a red card (either a heart or diamond) from the deck?

Probability of Opposite Events

Do the following problems about counting principle:

1) If you flip a coin, what is the probability that it will not land on tails?
2) A standard dice has six faces. What is the probability of **not** rolling a 3?
3) In a deck of 52 playing cards, there are 4 aces. What is the probability of **not** drawing an ace from the deck?
4) A jar contains 10 red balls, 8 blue balls, and 12 green balls. If you randomly choose one ball from the jar, what is the probability that you do **not** choose a blue ball?

www.mathnotion.com

5) In a class of 30 students, 18 are girls and 12 are boys. If a student is chosen at random, what is the probability that the student is **not** a girl?

Word Problems

Do the following word problems about statistics and probability:

1) The test scores of five students are: 85, 92, 76, 88, and 95. What is the mean score of the students?
2) If the average score of a students' 5 subjects is 84 and the scores of 4 subjects are 90, 73, 65, and 89, what is the score of the fifth subject?
3) A family's monthly expenses for groceries are: $150, $200, $180, $190, $160. What is the range of expenses?
4) If you flip two coins, how many possible outcomes are there?
5) You have 5 books, and you want to arrange them on a shelf. How many ways can you arrange the books?
6) How many ways can you arrange the letters in the word "MATH"?
7) You have 8 students in a class and need to choose 2 to help you with a project. How many ways can you choose 2 students?
8) A coin is flipped, and a six-sided die is rolled. What is the probability of getting a head on the coin and a number greater than 4 on the die?
9) In a class election, there are 3 candidates running for president and 4 candidates running for vice president. If you randomly select a president and a vice president, what is the probability of selecting Candidate A for president and Candidate B for vice president?
10) A bag contains 3 red balls, 4 green balls, and 5 blue balls. If two balls are drawn at random without replacement, what is the probability that both balls drawn are green?

Answer of Worksheets

Mean and Median

1) $\frac{40}{6}$
2) 24
3) approximately 143
4) 3.5
5) 1
6) 19
7) 28
8) 39
9) 4.65
10) 1.75

Mode and Range

1) 3
2) 14
3) 1
4) 30
5) 3.5
6) 120
7) 14
8) 22
9) 21
10) 25

Histograms

1)

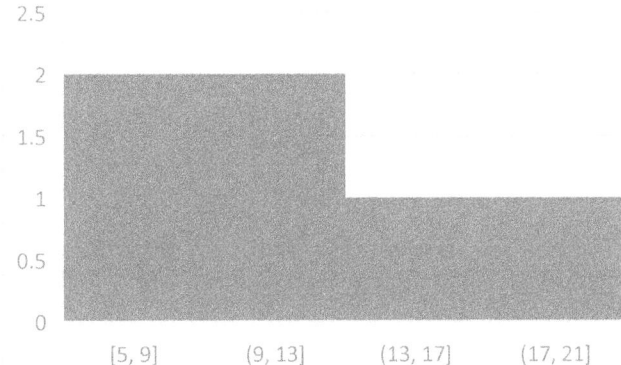

2)

3)

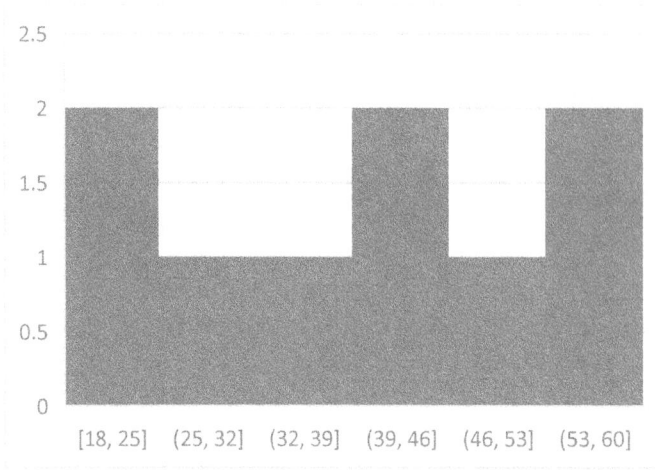

4) Highest frequency = [50, 79], Lowest frequency = [21, 50]
5) 41

Times Series

1)

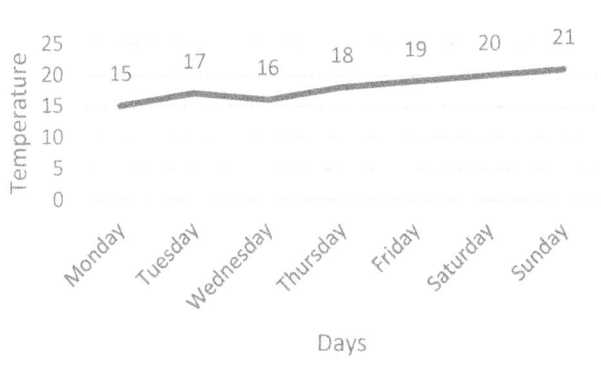

2)

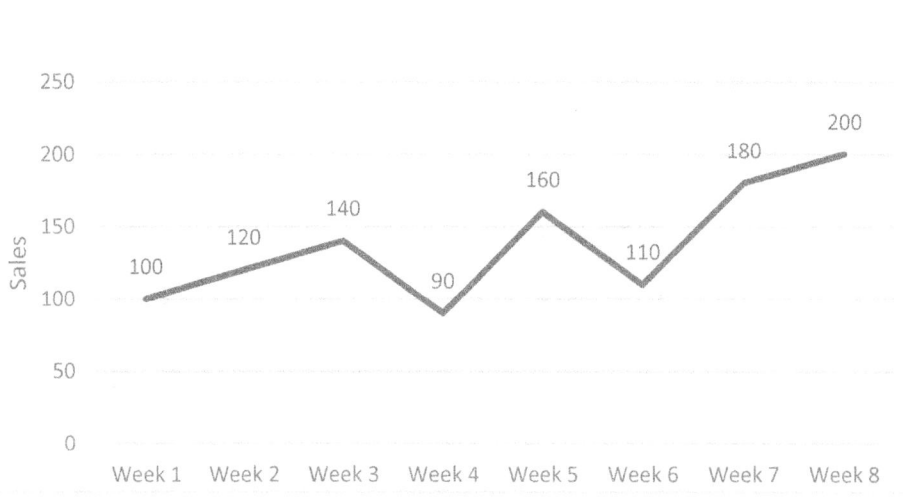

3)

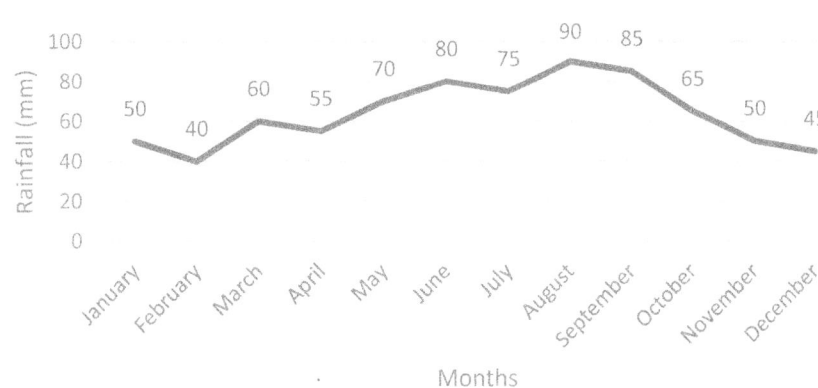

4) Most growth: between weeks 5 and 6, 6 and 7 and 9 and 10.
   Least growth: between weeks 1 and 2.
5) 13 mm.

Stem-and-Leaf Plot

1)
Stem	Leaf
1	2, 5, 8
2	2, 5, 9
3	0, 1

2)
Stem	Leaf
4	3, 5, 7
5	2, 6, 9
6	0, 0, 3

3)
Stem	Leaf
9	3, 5, 5
10	2, 4, 6
11	3, 4, 8, 9, 9

4)
Stem	Leaf
12	0, 6
13	1, 1, 9, 9
14	2, 5

5)
Stem	Leaf
25	4, 6, 9
26	2, 5, 8
27	3, 6, 9

Quartile of a Data Set

1) Median (Q2) = 15.5, Q1= 8, Q3= 23
2) Median (Q2) = 26, Q1= 18, Q3= 35
3) Median (Q2) = 27, Q1= 11.5, Q3= 40
4) Median (Q2) = 22, Q1= 13, Q3= 29
5) Median (Q2) = 36, Q1= 26.5, Q3= 45

Box-and- Whisker Plots

1) Minimum=5, Q1=7, Q2=9, Q3=12, Maximum=14

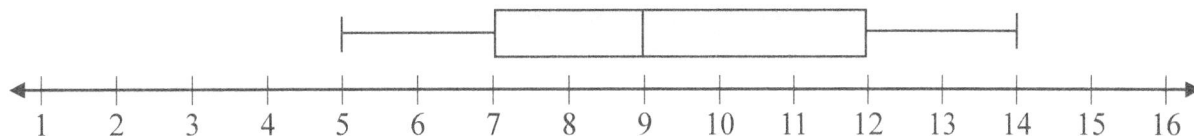

2) Minimum=15, Q1=20, Q2= 29, Q3=35.5, Maximum=40

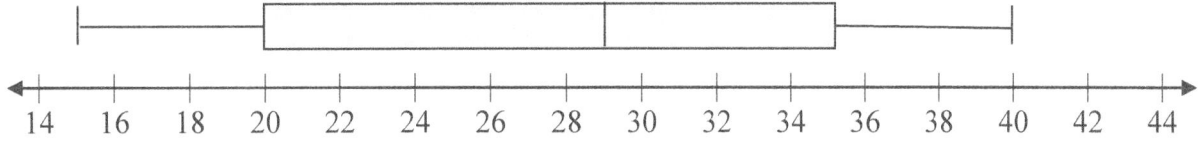

3) Minimum=28, Q1= 35, Q2= 41, Q3= 48, Maximum=52

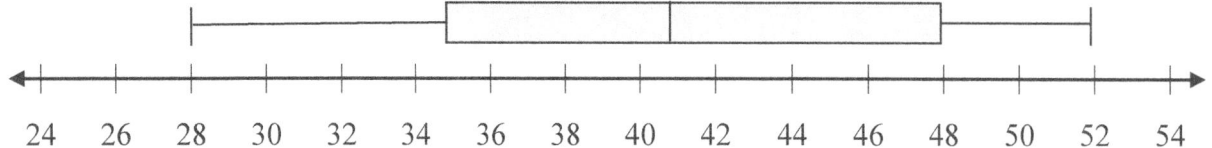

4) Minimum= 55, Q1=66.5, Q2=76.5, Q3= 83.5, Maximum= 90

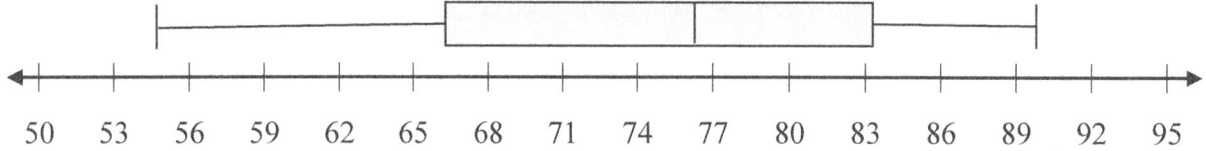

5) Minimum=102, Q1= 118, Q2= 132.5, Q3= 145, Maximum=160

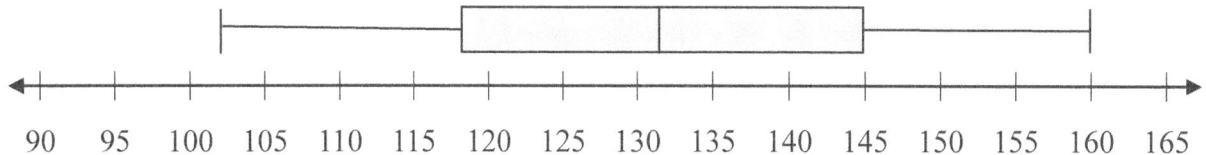

Dot Plots

1)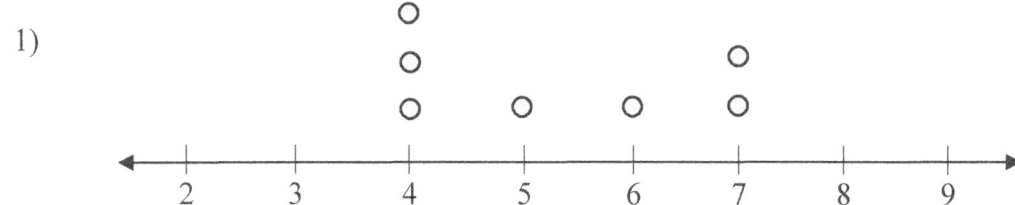

2)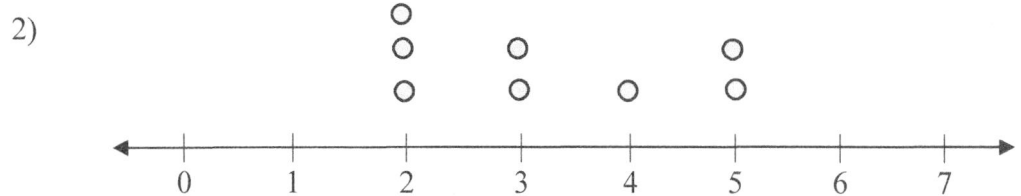

3)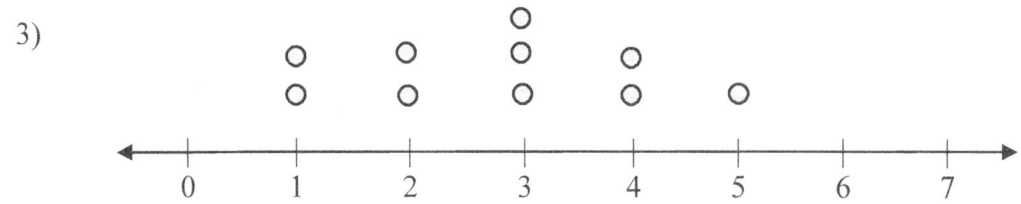

4)

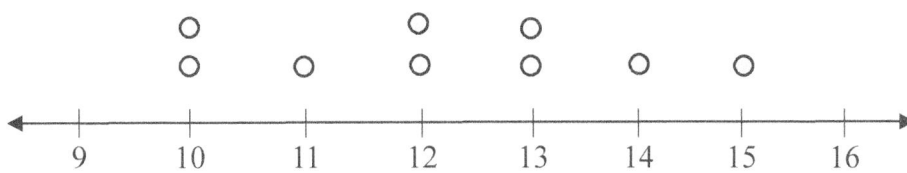

5)

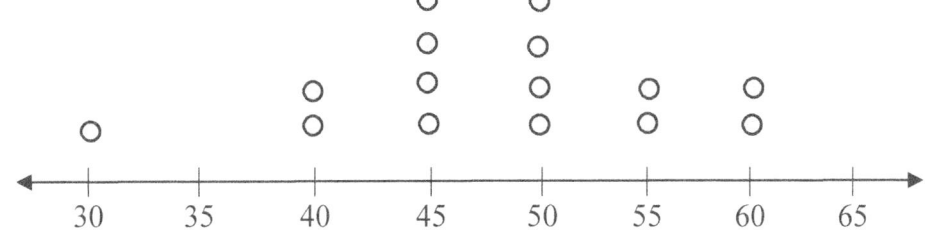

Pie Graph

1) $2,000
2) 5%
3) Fvorite fruit of students

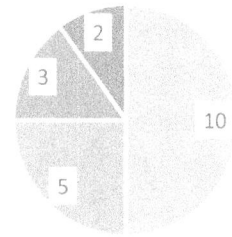

Apples ■ Bananas ■ Oranges ■ Grapes

4) Pets owned by students

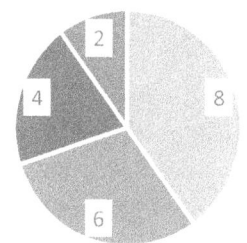

Dogs ■ Cats ■ Fish ■ Birds

5) Time spent on different activities

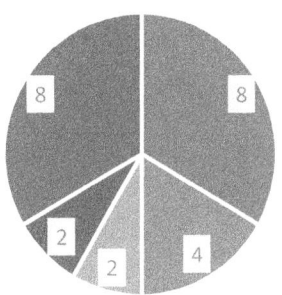

■ Sleeping ■ Studying ■ Playing ■ Watching ■ Other

Counting Principle

1) 9
2) 12
3) 16
4) 24
5) 24

Probability of Simple Events

1) $\frac{1}{6}$
2) $\frac{2}{5}$
3) $\frac{1}{13}$
4) $\frac{3}{10}$
5) $\frac{1}{2}$

Probability of Opposite Events

1) $\frac{1}{2}$
2) $\frac{5}{6}$
3) $\frac{12}{13}$
4) $\frac{11}{15}$
5) $\frac{2}{5}$

Word Problems

1) 87.2
2) 103
3) 50
4) 4
5) 120
6) 24
7) 28
8) $\frac{1}{6}$
9) $\frac{1}{12}$
10) $\frac{1}{11}$

Chapter 14: Practice Test

Time to Test

Time to refine your skill with a practice examination.

Take a REAL SBAC Mathematics test to simulate the test day experience. After you have finished, score your test using the answer key.

Before You Start

- You will need a pencil and scratch paper to take the test.
- For this practice test, do not time yourself. Spend as much time as you need.
- It is okay to guess. You will not lose any points if you're wrong.
- **Calculators are not permitted for sixth grade SBAC Tests.**

After you have finished the test, review the answer key to see where you went wrong.

Good Luck!

Formula Sheet

SBAC Mathematics Formula Sheet Grade 6

Shape	Area	Perimeter
Triangle	$A = \frac{1}{2}bh$	
Rectangle	$A = lw$	$P = 2l + 2w = 2(l+w)$
Trapezoid	$A = \frac{1}{2}h(b_1 + b_2)$	
Parallelogram	$A = bh$	
Square	$A = s \times s$	

Key	
b = base	l = length
h = height	w = width
B = area of base	s = side length
H = height of triangular prism	
s_1, s_2, s_3 are the lengths of each side of the triangular base	

3 – Dimensional Shape	Volume	Surface Area
Rectangular Prism	$V = lwh = Bh$	$SA = 2lw + 2lh + 2wh = 2B + 2lh + 2wh$
Triangular Prism		$SA = bh + (s_1 + s_2 + s_3)H = 2B + (s_1 + s_2 + s_3)H$

Standard Units	Metric Units
Conversions – Length	
1 yard (yd) = 3 feet (ft) = 36 inches (in.)	1 meter (m) = 100 centimeters (cm)
1 mile (mi) = 1,760 yards (yd) = 5,280 feet (ft)	1 meter (m) = 1,000 millimeters (mm)
	1 kilometer (km) = 1,000 meters (m)
Conversions – Volume	
1 cup (c) = 8 fluid ounces (fl. oz.)	1 liter (l) = 1000 milliliters (ml)
1 pint (pt) = 2 cups (c)	1 liter (l) = 1,000 cubic centimeters (cu. cm)
1 gallon (gal) = 128 fluid ounces (fl. oz.)	
1 gallon (gal) = 4 quarts (qt)	
Conversions – Weight/Mass	
1 pound (lb) = 16 ounces (oz)	1 gram (g) = 1000 milligrams (mg)
1 ton (T) = 2000 pounds (lb)	1 kilogram (kg) = 1000 grams (g)

SBAC Practice Test

1. What is the value of (4^3)?
   A. 12
   B. 16
   C. 64
   D. 81

2. What is the absolute value of -12?
   A. -12
   B. 0
   C. 12
   D. 24

3. Which of the following numbers is a prime number?
   A. 15
   B. 23
   C. 27
   D. 33

4. What is the value of ($\frac{3}{5}$) as a decimal?
   A) 0.3
   B) 0.35
   C) 0.6
   D) 0.65

5. What is the perimeter of a square with side length 9 cm?
   A) 18 cm
   B) 27 cm
   C) 36 cm
   D) 81 cm

6. What is the value of ($5 + 3 \times 2$)?
   A) 10
   B) 11
   C) 13
   D) 16

7. What is the area of a rectangle with a length of 12 cm and a width of 7 cm?
   A) 19 cm²
   B) 38 cm²
   C) 72 cm²
   D) 84 cm²

8. What is the least common multiple (LCM) of 8 and 12?

   A) 12

   B) 24

   C) 36

   D) 48

9. What is the value of $(\frac{7}{8} - \frac{3}{8})$?

   A) $\frac{1}{2}$

   B) $\frac{1}{4}$

   C) $\frac{3}{8}$

   D) $\frac{5}{8}$

10. What is the greatest common factor (GCF) of 36 and 48?

    A) 6

    B) 12

    C) 18

    D) 24

11. What is the median of the following data set? [12, 15, 18, 20, 22, 25, 30]

    A) 18

    B) 20

    C) 22

    D) 25

12. What is the mode of the following data set? [5, 7, 7, 8, 9, 9, 9, 10]

    A) 5

    B) 7

    C) 9

    D) 10

13. Solve for (x): ($2x - 7 = 11$).

    A) 2

    B) 5

    C) 9

    D) 18

14. Simplify: ($3(4 + 5) - 6$).

    A) 15

    B) 21

    C) 27

    D) 33

15. What is the area of a triangle with a base of 10 cm and a height of 8 cm?

    A) 18 cm²

    B) 40 cm²

    C) 80 cm²

    D) 160 cm²

16. What is the volume of a rectangular prism with a length of 5 cm, width 3 cm, and height 4 cm?

    A) 12 cm³

    B) 60 cm³

    C) 120 cm³

    D) 180 cm³

17. What is the circumference of a circle with a radius of 7 cm? (Use $\pi = 3.14$)

    A) 21.98 cm

    B) 43.96 cm

    C) 153.86 cm

    D) 307.72 cm

18. What is the area of a circle with a radius of 5 cm? (Use $\pi = 3.14$)

    A) 15.7 cm²

    B) 31.4 cm²

    C) 78.5 cm²

    D) 157 cm²

19. What is the surface area of a cube with side length 4 cm?

    A) 16 cm²

    B) 64 cm²

    C) 96 cm²

    D) 128 cm²

20. A recipe requires $\frac{3}{4}$ cup of sugar. If you want to make 6 batches of the recipe, how many cups of sugar will you need?

    A) $4\frac{1}{2}$

    B) $5\frac{1}{4}$

    C) $6\frac{3}{4}$

    D) $7\frac{1}{2}$

21. If you save $50 each month for a year, how much will you have saved at the end of the year?

    A) $500

    B) $600

    C) $700

    D) $800

22. A store offers a 20% discount on a $50 item. What is the sale price?

   A) $10

   B) $30

   C) $40

   D) $45

23. If 3 pounds of apples cost $6, how much will 7 pounds of apples cost?

   A) $10

   B) $12

   C) $14

   D) $16

24. A car travels 240 miles in 4 hours. What is its speed?

   A) 40 mph

   B) 50 mph

   C) 60 mph

   D) 70 mph

25. Which of the following numbers is both a perfect square and a perfect cube?

   A) 16

   B) 64

   C) 100

   D) 125

26. What is the unit rate for 15 ounces of juice costing $4.50?

   A) $0.20 per ounce

   B) $0.25 per ounce

   C) $0.35 per ounce

   D) $0.30 per ounce

27. What is the value of $\frac{5}{8} \div \frac{1}{4}$?

   A) $\frac{5}{32}$

   B) $\frac{5}{2}$

   C) $\frac{5}{4}$

   D) $\frac{5}{8}$

28. Solve for y: $\frac{y}{3} = 9$).

   A) 3

   B) 12

   C) 27

   D) 36

29. What is the product of (7 × (−8))?

    A) −56

    B) −15

    C) 15

    D) 56

30. Simplify: (4x + 3 − 2x + 5).

    A) (2x + 8)

    B) (6x + 8)

    C) (2x − 2)

    D) (6x − 2)

31. If 6 workers can complete a job in 8 days, how many workers are needed to complete the same job in 4 days?

    A) 8

    B) 10

    C) 12

    D) 16

32. What is the unit rate for 12 ounces of juice costing $3?

    A) $0.20 per ounce

    B) $0.25 per ounce

    C) $0.30 per ounce

    D) $0.35 per ounce

33. If 5 pencils cost $1.50, how much will 12 pencils cost?

    A) $2.50

    B) $3.00

    C) $3.60

    D) $4.50

34. What is the value of (2^5)?

    A) 10

    B) 16

    C) 32

    D) 64

35. What is the area of a parallelogram with a base of 12 cm and a height of 5 cm?

    A) 17 cm²

    B) 30 cm²

    C) 60 cm²

    D) 120 cm²

36. What is the value of $\frac{3}{4} \times \frac{2}{5}$?

    A) $\frac{1}{10}$

    B) $\frac{3}{10}$

    C) $\frac{5}{10}$

    D) $\frac{6}{10}$

Answer Key

❋ Now, it is time to review your results to see where you went wrong and what areas you need to improve!

Practice Test			
1	C	19	C
2	C	20	A
3	B	21	B
4	C	22	C
5	C	23	C
6	B	24	C
7	D	25	B
8	B	26	D
9	A	27	B
10	B	28	C
11	B	29	A
12	C	30	A
13	C	31	C
14	B	32	B
15	B	33	C
16	B	34	C
17	B	35	C
18	C	36	B

Answers and Explanations

1) Answer: C) 64

(4^3) means ($4 \times 4 \times 4$), which equals 64.

2) Answer: C) 12

Absolute value is the distance from 0 on the number line, so ($|-12| = 12$).

3) Answer: B) 23

A prime number has no divisors other than 1 and itself. 23 is the only prime number in the list.

4) Answer: C) 0.6

Divide 3 by 5 to get 0.6.

5) Answer: C) 36 cm

Perimeter of a square = ($4 \times \text{side} = 4 \times 9 = 36$) cm.

6) Answer: B) 11

Follow the order of operations (PEMDAS): ($3 \times 2 = 6$), then ($5 + 6 = 11$).

7) Answer: D) 84 cm²

Area = length × width = ($12 \times 7 = 84$) cm².

8) Answer: B) 24

The LCM is the smallest number that both 8 and 12 divide into evenly. The multiples of 8 are 8, 16, 24, and the multiples of 12 are 12, 24. So, the LCM is 24.

9) Answer: A) $\frac{1}{2}$

Subtract the numerators: $\frac{7-3}{8} = \frac{4}{8} = \frac{1}{2}$.

10) Answer: B) 12

The GCF is the largest number that divides both 36 and 48 without a remainder. The factors 36 are 1, 2, 3, 4, 6, 9, 12, 18, 36, and the factors of 48 are 1, 2, 3, 4, 6, 8, 12, 16, 24, 48. The largest common factor is 12.

11) Answer: B) 20

The median is the middle number in an ordered list. Here, 20 is the middle number.

12) Answer: C) 9

The mode is the number that appears most often. Here, 9 appears three times.

www.mathnotion.com

13) Answer: C) 9

Add 7 to both sides: ($2x = 18$). Then divide by 2: ($x = 9$).

14) Answer: B) 21

Use the order of operations (PEMDAS): ($3(9) - 6 = 27 - 6 = 21$).

15) Answer: B) 40 cm²

Area = $\frac{1}{2}$ base × height = $\frac{1}{2} \times 10 \times 8 = 40$ cm².

16) Answer: B) 60 cm³

Volume = length × width × height = ($5 \times 3 \times 4 = 60$) cm³.

17) Answer: B) 43.96 cm

Circumference = $2\pi r = 2 \times 3.14 \times 7 = 43.96$.

18) Answer: C) 78.5 cm²

Area = $\pi r^2 = 3.14 \times 5^2 = 78.5$ cm².

19) Answer: C) 96 cm²

Surface area = $6 \times side^2 = 6 \times 16 = 96$ cm².

20) Answer: A) $4\frac{1}{2}$

Multiply $\frac{3}{4}$ by 6: $\frac{3}{4} \times 6 = \frac{18}{4} = 4\frac{1}{2}$.

21) Answer: B) $600

Multiply the monthly savings by 12: $50 \times 12 = 600$.

22) Answer: C) $40

A 20% discount means you pay 80% of the original price: $0.80 \times 50 = 40$.

23) Answer: C) $14

First, find the cost per pound: $\frac{6}{3} = 2$ dollars per pound. Then multiply by 7: $7 \times 2 = 14$.

24) Answer: C) 60 mph

Speed = distance ÷ time = $\frac{240}{4} = 60$ mph.

25) Answer: B) 64

A perfect square is a number like 16 (4^2) or 64 (8^2). A perfect cube is a number like 64 (4^3) or 125 (5^3). 64 is both 8^2 and 4^3.

26) Answer: D) 0.30

Unit rate = total cost ÷ quantity = $\frac{4.5}{15} = 0.30$ dollars per ounce.

27) Answer: B) $\frac{5}{2}$

Dividing by a fraction is the same as multiplying by its reciprocal: $\frac{5}{8} \times \frac{4}{1} = \frac{20}{8} = \frac{5}{2}$.

28) Answer: C) 27

Multiply both sides by 3: $y = 27$.

29) Answer: A) -56

A positive number multiplied by a negative number is negative: $7 \times (-8) = -56$.

30) Answer: A) $(2x + 8)$

Combine like terms: $4x - 2x = 2x$ and $3 + 5 = 8$.

31) Answer: C) 12

This is an inverse proportional problem. $6 \times 8 = x \times 4$, so $x = 12$

32) Answer: B) $0.25 per ounce

Unit rate = total cost ÷ quantity = $\frac{3}{12} = 0.25$ dollars per ounce.

33) Answer: C) $3.60

First, find the cost per pencil: $\frac{1.50}{5} = 0.30$ dollars per pencil. Then multiply by 12: $0.30 \times 12 = 3.60$.

34) Answer: C) 32

$2^5 = 2 \times 2 \times 2 \times 2 \times 2 = 32)$.

35) Answer: C) 60 cm²

Area = base × height = $12 \times 5 = 60$ cm².

36) Answer: B) $\frac{3}{10}$

Multiply the numerators and denominators: $\frac{3 \times 2}{4 \times 5} = \frac{6}{20} = \frac{3}{10}$.

www.ingramcontent.com/pod-product-compliance
Lightning Source LLC
Chambersburg PA
CBHW082202070526
44585CB00020B/2251